Consider a Cylindrical Cow

More Adventures in Environmental Problem Solving

Consider a Cylindrical Cow

More Adventures in Environmental Problem Solving

John Harte
University of California, Berkeley

University Science Books
Sausalito, California

University Science Books
www.uscibooks.com
Order Information:
Phone (703) 661-1572
Fax (703) 661-1501

Production manager: *Susanna Tadlock*
Design consultant: *Robert Ishi*
Compositor: *Asco Typesetters*
Printer & binder: *Edwards Brothers, Inc.*

This book is printed on acid-free paper.

Library of Congress Cataloging-in-Publication Data

Harte, John, 1939-
Consider a cylindrical cow: more adventures in environmental problem solving / John Harte.
p. cm.
Includes bibliographical references and index.
ISBN 1-891389-17-3 (alk. paper)
1. Environmental sciences —Mathematics. 2. Environmental sciences—Problems, exercises, etc. I. Title.

GE45.M38 H37 2000
363.7'001'51—dc21 00-047663

Printed in the United States of America
10 9 8 7 6 5 4 3 2 1

For Ethan and Daniel

Contents

Preface

The cow is back. This time she is cylindrical, not spherical. Still no legs, udders, or head, but the torso is a more realistic shape.

Following on the hooves of *Consider a Spherical Cow* (Harte, J. 1988. *Consider a spherical cow: a course in environmental problem solving.* Sausalito, CA: University Science Books) called *COW-1* herein, *COW-2* will teach you additional modeling skills of use in environmental science. Some of these skills are at roughly the same mathematical level as Chapters II and III of *COW-1*, whereas others are more advanced. Although the emphasis here, as in *COW-1*, is on analytic approaches to squeezing information from models, I also include spread-sheet methods for simulating the behavior of models constructed from differential equations. You will get the most out of *COW-2* if you have worked through *COW-1*, but I have tried to make *COW-2* self-contained.

I have assumed that you, the reader, are acquainted with both differential and integral calculus, though the relationship need not be lustful. Indeed, only a passionate dislike (or deep and unyielding fear) of mathematics will disqualify you from proceeding further. Some past exposure to matrices will also help; for those who need to do some remedial reading on calculus or matrix algebra, I highly recommend Clifford Swartz's excellent *Used Math* (Swarts, C. E. 1993. *Used math: for the first two years of college science.* College Park, MD: AAPT Press).

I have also assumed that you are ready to seek out deeper insights into ways of modeling the complexities of nature, but the underlying goal is the same as in *COW-1*: to teach ways of stripping away inessential detail and capturing with mathematics the essentials of a complex system.

Why do I want you to learn to use mathematics? Can't we just *talk* about environmental problems? There is another reason in addition to the often-expressed, and correct, argument that mathematics seems to be the "language that nature speaks" and therefore facilitates the understanding of nature. Mathematics is a kind of global language that not only needs no translation from nation to nation but also bridges the disciplines. Economists and scientists often confuse each other (and sometimes themselves) using everyday language to describe how things depend upon other things; the reason is that in the two academic traditions the same word may be used in different senses. (Consider, for example, the terms equity, parity, derivative, and stock.) Mathematics, however, has a way of cutting through such confusion.

Just as the "Tools of the Trade" chapter of *COW-1* was structured around the topics of steady-state box models, thermodynamics, chemical equilibrium theory, and non-steady-state box models, *COW-2* is structured around the central themes of probability, optimization, scaling, differential equations, stability, and feedback. Why these themes? Consider such problems as (a) determining how some shift in land-use practice (e.g., conversion of tropical forest to grazing land) might result in species extinction or climate alteration in the region, or (b) determining whether exposure to a trace substance released from a factory poses a serious health threat to people living nearby. In problems such as these, themes that affect our understanding and analysis of the problem can be extracted as follows:

1. Much of our information about these situations comes to us in the form of data derived from some sampling scheme. We need to know something about probability and statistics to assess how representative the data are.

2. We may discover that the activity leading to the environmental damage does bring some benefit to society, but that if the activity is too intense, then society is the loser. We need to know something about optimization to estimate where the balance point between societal benefits and societal costs lies.

3. We may be able to determine something about the effects of a toxic substance on adults but not on children, so we need to know something about how to scale our knowledge from big people to little people. Or, to estimate species loss, we may be able to use existing data on the rate of local species extinctions from deforestation, provided, we can "scale up" to larger regions.

4. We will want to predict not only the eventual long-term-averaged climate conditions that will result from the loss of forest but also the way in which climatic conditions will change over time in the shorter term as they approach that eventual state. The mathematical tools needed to predict that climate trajectory over time are differential equations. We need to know how to set up the appropriate differential equation for the problem at hand and how to find approximate solutions to that equation.

5. As we alter a complex system, for example by cutting down trees, thresholds of instability may be crossed. Before the threshold is reached, the system can assimilate the stress and maintain some sort of equlibrium, but if the stress is too great, then the system may undergo

a dramatic response. We need to know how to estimate whether such a thereshold exists and where it lies.

6. Climate changes induced by altered land-use practices can further alter ecosystems and degrade ecosysystem services of benefit to society. These feedbacks may further alter human behavior as people seek to compensate for the lost ecosystem services. We need to be able to estimate the magnitude and consequences of such feedback effects.

The pervasiveness of these themes is well known to environmental scientists in fields such as climatology, ecology, hydrology, toxicology, and atmospheric chemistry. Despite this understanding among scientists, public policy debates often neglect such critical linkages and issues. The public is easily misled by twisted probabilistic reasoning. Misuse of optimization methods can lead to suboptimization (doing better and better something that would not have been done at all if the boundary of analysis had been enlarged). Analytic methods, insights, institutions, adaptations, and solutions appropriate at one scale (e.g., experimental plots in ecology, individual firms or power plants in industry, a grid square in global climate modeling) are often naively and mistakenly extrapolated to other scales. Incorrect conclusions about the distribution and fate of emitted substances usually can be spotted and corrected with the aid of approximation methods for studying solutions to differential equations. Instabilities and feedbacks, both between and within human and biophysical systems, are often ignored in political, social, and even scientific analysis, leading society to be inadequately prepared for the possible abruptness and intensity of resulting changes.

At the beginning of each of the five thematic sections in this book is a general treatment of the relevant mathematical concepts. The core of the book is 25 fully worked-out problems. As in *COW-1*, each problem statement is posed more like a research question than a standard homework exercise. Homework exercises follow each worked-out problem solution and range from relatively straightforward exercises that probe readers' understanding of the concepts and methods to more difficult and open-ended research suggestions. The Appendix summarizes many useful mathematical notations and formulae you are likely to encounter.

Acknowledgments

I was fortunate to be able to write this book at two most congenial research institutions: the Centre for Population Biology at Silwood Park, UK, where I spent a sabbatical, and the Rocky Mountain Biological Laboratory in Colorado, where I regularly conduct summer field research in ecology. Financial support from the U.K. National Environmental Research Council and the Class of 1935 Endowed Chair at the University of California, Berkeley, is gratefully acknowledged.

Many colleagues have contributed, through their writings and their conversations with me, to the clarification of the problems presented in this book. I especially wish to thank Steven Fetter, Richard Garwin, Jessica Green, Ken Harte, Dan Kammen, Ann Kinzig, Jim Kirchner, Harvey Michaels, Zack Powell, Robert Socolow, and Margaret Torn. I am also very grateful to Peter Armbruster, Christopher Buck, Jessica Green, Mary Ellen Harte, Ann Kinzig, and Paul Bunje for thoughtful comments on a draft version of the book. Working with the production staff at University Science Books has again been a pleasure; I particularly want to thank Bruce Armbruster and Jane Ellis for their wise guidance, Ann McGuire for a fine copy edit, Bob Ishi for his splendid designs, and Susanna Tadlock for easing the book painlessly through the labors of production.

I am once again indebted to my wife, Mary Ellen, for her skill in drawing cartoons and illustrating technical points. With "Mel" and our daughter, Julia, I enjoyed many utterly delightful respites from writing, which ultimately made it all worthwhile.

Chapter I
Probability

> *Gonzalo: "Our hint of woe is common; every day some sailor's wife, the master of some merchant, and the merchant, have just our theme of woe. But for the miracle, I mean our preservation, few in millions can speak like us."*
>
> —*William Shakespeare*, The Tempest

Introduction

No branch of mathematics is more widely used and misused in everyday life than probability. Even when risks to our own health are involved, our assessments at times defy logic. The problem generally lies not so much in ignorance of the rules of probability but rather in a failure to know how to apply the rules to the particular circumstances at hand. I often confront a similar failure in students in my environmental science classes: in response to a challenging quantitative homework problem, they assert that they do not know the mathematics needed to solve the problem. In fact, the only mathematics needed may just be simple multiplication. The students know that branch of mathematics perfectly well! They just don't know *when* they need to multiply and *what* they need to multiply by what.

Gonzalo is not alone in making muddled pronouncements about risk and probability. To illustrate common misuses of probabilistic thinking about risk, here is a sampling of some typical pronouncements culled and paraphrased from the media, from acquaintances, and in one case from a paper in a scientific journal (they probably sound familiar):

"The chances of dying because of a major accident at a nuclear power plant are far less than the chances of being killed in a big earthquake in California.... Residents of that state need not be worried about nuclear power."

"Look what happened when Harry went off his organic food diet for just one day—he got a terrible migraine later that week. You're crazy to shop at the supermarket."

"Far more people get killed in car crashes than from using this chemical spray—why should I stop using it?"

"Rat studies show that many foods contain natural cancer-causing chemicals that are more potent than the pesticides in our diet—we should stop asking the government to waste taxpayers' money regulating pesticides."

"Aunt Sheila smoked all her adult life and she lived to the age of 100—why should I worry?"

"The chance of being attacked by a shark is very small, and so the chance of being attacked twice is much smaller still. So, because I survived a shark attack while surfing last summer, I can now go back to surfing reassured that the chance is negligible that I will be attacked again."

Much of the confusion about risk and probability implicit in these comments is generated by the failure of the media, and scientists as well, to level with the public about the nature of risk, the limits of science, and the inevitable role of personal values. People tend to ignore the fact that life is intrinsically a gamble. People respond medically in different ways to the same toxic stress, and the most that risk studies can do is tell you the chances that harm will occur to the average person drinking from a certain water supply or living near a particular factory. The probability of harm to an *individual* will be influenced by the individual's behavior, of course. For example, the same can of paint remover can be used by two different people in very different ways; one person may receive a much larger exposure than the other because of careless handling of the product. Even widely cited estimates of risk to the average person are often highly uncertain; many cancer-risk studies are done using test animals and unrealistically high doses of the chemical under suspicion, leading to very uncertain estimates of human risk at more commonly encountered lower doses.

Science cannot tell you how much risk you should be willing to accept as a consequence of your daily actions, nor can it provide a widely acceptable common scale on which to compare risks. You, and only you, must decide whether the chance of some horrible thing happening is so great that you should not eat this or live near that. A further complexity, illustrated by some of the quotes above, is that with every choice you make among alternative courses of action there are risks and benefits; comparing only the risks (which is what most of the publicity is about) may be foolish, for the choice that carries the greatest risk may also lead to the greatest benefit. Moreover, there is no scientific way to determine which of the following risks is more acceptable: an extremely low chance of a catastrophic event, or a much higher chance of a moderately unpleasant event. Also, risks differ sharply in the degree to which you have control over your exposure.

Thus, some risks are self-imposed (e.g., sky diving); others are somewhat under your control but still partially unavoidable (e.g., you could move away from a polluted city if you had the money and a job elsewhere); still others are almost completely out of your control (e.g., the risk of global nuclear war). Science cannot tell you how to take into account these different qualitative features of risks.

You may come away from this discussion shattered by the disturbing "grayness" of choices concerning risk and distrustful of any scientific results about risk. That would be a grave mistake, for science can certainly help illuminate the debates. Laboratory studies, for example, do shed light on the probability of a toxic exposure causing an illness in test animals; statistical studies can be used to determine whether human populations exposed to certain toxics are more prone to certain illnesses. But the results of such studies will never provide the whole picture, and those who claim they will sow the seeds of confusion, not enlightenment. In the end, science can supply us only with information not values. For evaluating information and deciding how to act upon it, there still is not, and never will be, any substitute for common sense.

The problems that follow in this book are designed to enrich your understanding of probabilistic thinking; they will not turn you into a more sensible person, but they will provide some tools to help you express your native common sense more precisely.

EXERCISE 1: Examine carefully each of the six quotes above, and categorize in generic terms the types of confusion exemplified by each.

Background

You need to know only a few basic rules to be a successful gambler or bookie, or a probabilistically proficient citizen. These rules concern how you estimate the probability of something happening and how you amalgamate probabilities for various events to occur when you know the probability of each event separately. To understand the origin and applicability of these rules, consider two different types of situations in which probabilities play a role: one that involves *mutually exclusive* events and a second that involves *independent* events.

Mutually exclusive events are exemplified by the classic situation of blindly withdrawing, one at a time, red, green, and blue balls from a bag containing some of each. The choices are mutually exclusive in the sense that selection of one of the colors precludes selection of the other two in any particular draw. The three outcomes of a selection, a red or a blue or a green ball, are a complete set of outcomes in the sense that one of those three must result when a ball is selected. If the probabilities of choosing the three colors on a particular draw are p_r, p_b, and p_g, then the sum of the probabilities equals 1: $p_r + p_b + p_g = 1$. More generally, Rule 1 can be stated as follows:

RULE 1: *The sum of the probabilities of a complete set of mutually exclusive events must equal 1.*

An important corollary to Rule 1 is that if p is the probability of an event occurring, then $1 - p$ is the probability of it not occurring. Note that $1 - p_r - p_b$ is, by Rule 1, equal to p_g. But p_g is also equal to 1 minus the probability of the selected ball not being green. Hence, $p_r + p_b$ equals the probability of the selected ball not being green. More generally, we can state the corollary as follows: the sum of the probabilities of a subset of mutually exclusive events is equal to the probability of the remaining events not occurring. Rule 1 and its corollary formalize the important, albeit self-evident, assumptions of completeness and mutual exclusivity.

In the second type of situation, a set of independent, rather than mutually exclusive, events can occur. What is the chance of drawing the same-color ball twice in a row? To answer this, let's back up a minute and ask how the probabilities, p_r, p_b, p_c, are determined. Assuming the balls are well mixed, then each probability is just the ratio of the number (N) of balls of that color divided by the total number of balls. For example, $p_r = N_r/(N_r + N_b + N_g)$. After the first draw, however, the number of balls has changed, and so the probabilities will change as well. That is another way of saying that the events are not independent—choosing a red ball the first time influences the chance of choosing a red ball the second time. To avoid having to deal with this correction, and to make the probabilities describing a sequence of draws truly independent of each other, we will assume that the number of balls of each color is so large that the change in abundance and thus the changes in the probabilities after the first draw can be ignored. Equivalently, we can assume that balls are placed back in the bag after they are drawn out. Then the events can be considered independent of one another, and the original assignment of probabilities can be used at each draw. Consider two sequential draws from the bag of colored balls. Each time, the chance of picking a red ball is p_r, and so the chance of doing that twice in a row is $(p_r)^2$. The probability of first picking a red ball and then a green ball is $p_r p_g$. Rule 2 is as obvious as Rule 1:

RULE 2: *The probability that a set of independent events will occur is the product of the chances that each will occur.*

Rule 2 applies to the situation in which the events in the set are independent of one another, but the individual probabilities of some events depend upon whether or not other events occur. In that case, we have a modification of Rule 2:

RULE 2′: *The probability, $P(A, B)$, that two events, A and B, will occur is $P(A)P(B|A)$, where $P(B|A)$ is the conditional probability that B occurs if A occurs.*

Note that if A and B are truly independent, then $P(B|A) = P(B)$, and we are back to Rule 2. Note, also, that Rule 2′ is not restricted to events that occur in the sequence of first A and then B; for example, the events might be different values of the abundance of a species in adjacent patches of habitat.

Using Rules 1 and 2, what is the chance that you will select at least one red ball in two draws? The draws are independent of each other, and the choices within each draw are mutually exclusive. Let's work that out two different ways. First, the chance is equal to 1 minus the probability of not getting a red ball in either draw. The probability of not getting a red ball in either draw is, by Rule 2, the product of the probability of not getting a red ball in each draw, or $(1 - p_r)^2$. Hence the chance of selecting at least one red ball in two draws is $\mathbf{1 - (1 - p_r)^2}$. Equivalently, by our corollary to Rule 1, the chance is equal to the sum: (probability of selecting a red ball in both draws) + (probability of selecting a red ball in the first but not the second draw) + (probability of selecting a red ball the second but not the first draw). The probability of getting red balls on both draws is p_r^2. The chance of getting a red ball the first but not the second or the second but not the first draw is $p_r(1 - p_r)$. Hence, the probability of selecting at least one red ball equals $p_r^2 + 2p_r(1 - p_r) = \mathbf{2p_r - p_r^2}$. You can readily show that these two bold-faced expressions are identical.

Here is a more complex situation that allows us the opportunity to put both rules into action. Imagine that you want to travel from town A to town D, first taking a taxi from A to B, then a plane from B to C, and then a bus from C to D. You are worried about not reaching your destination because taxi drivers, critical airline employees such as pilots and refuelers, and bus drivers each occasionally go on strike. Over the past 10 years, taxi drivers in town A have been on strike a fraction f_t of the time, critical airline people associated with your route have been on strike a fraction f_a of the time, and bus drivers working for the bus company that connects towns C and D have been on strike a fraction f_b of the time. Moreover, a strike in one transportation sector has no influence on whether there is a strike in either of the other two sectors (no sympathy strikes), which is equivalent to the statement that strikes are independent events.

What is the chance that a strike will prevent your trip from happening? Following the logic deployed above, let's first calculate the chance that no strike will occur and the trip will be completed. For the trip to occur successfully, three events must occur in combination: no taxi strike, no airline strike, and no bus strike. According to Rule 2, the chance that the three events will occur in combination is the product of the chance that each will occur. The chance there will be no taxi strike is, by Rule 1, 1 minus the chance there will be a strike, or $1 - f_t$. Similarly, $1 - f_a$ and $1 - f_b$ are the chances there will be no airline and bus strikes, respectively. Hence, the probability that you will not be strike-plagued is $(1 - f_t)(1 - f_a)(1 - f_b)$. Thus, the chance that a strike will disrupt your plans is $1 - (1 - f_t)(1 - f_a)(1 - f_b)$.

What is the chance that there will be a taxi strike and an airplane strike but not a bus strike? By combining our rules, the answer is $f_t f_a(1 - f_b)$; do you see why?

In the course of working out these examples, I have actually sneaked Rule 3 by you. I assumed above that f_t was the probability of a taxi strike. How could I do that? The f values are fractions of time; they tell us the portion of time over the past 10 years that a strike occurred. How can we equate the fraction of time that strikes have occurred in the past to the probability of a strike in the future? We do it in two steps. First, we equate the f values to the probability of a strike at any time over the past 10 years. This step is analogous to our statement that $N_r/(N_r + N_b + N_g)$ was the probability of drawing a red ball. There, we assumed that the balls in the bag were well mixed, so that whatever place in the bag you reached, the best estimate of the chance of picking a red ball was determined by the spatially averaged abundance. For the strike situation, we are assuming that, in effect, time is well mixed! In other words, the average probability equals the instantaneous probability. The second step consists of assuming that the same probability characterizes the future as characterized the past. Why is that assumption reasonable? The answer has to do with our ignorance. Knowing nothing but the information given in the problem statement, the safest assumption we can make is that the probability of a strike occurring on a particular day in the future equals the probability of it occurring on a particular day in the past. In a loose sense, the past and the future are well mixed. If we knew more information —perhaps that the pilots had just won a big salary increase—then our assumption that the future will, statistically speaking, look like the past would not be a good one. These assumptions about spatial and temporal uniformity are often called for in real-world situations. The reason is that the information we have access to is often not the probabilities themselves but rather indirect information from which we can estimate probabilities only by making such assumptions. Now we can state Rule 3, which pertains to the estimation of probabilities from empirical information.

RULE 3: *In the absence of additional information, the probability of the future occurrence of events is best estimated by the frequency of their past occurrence, and the probability of the appearance of objects is best estimated by their relative abundance.*

Given a set of N numbers, labeled n_i where i ranges from 1 to N, their average or mean value, which we denote by $\langle n \rangle$, is given by the familiar rule:

RULE 4:

$$\langle n \rangle = \left(\frac{1}{N}\right) \sum_{i=1}^{N} n_i,$$

where the symbol $\sum$ means "take the sum," and the sum is over all values of i from 1 to N.

Next, we formulate an important rule that allows you to estimate the average outcome of a repeated sequence of events governed by probabilities. Suppose that red balls are worth 1 point, blue are worth 2 points, and green are worth 3 points. How many points would you expect to have earned after a large number of draws? A fraction p_r of the draws will result in 1 point, a fraction p_b will result in 2 points, and a fraction p_g will result in 3 points. So the expected number of points per draw is $(1 \times p_r) + (2 \times p_b) + (3 \times p_g)$. More generally,

RULE 4′: *expected value of the mean* = $\sum$ *(value of outcome) (probability of outcome).*

Here the sum is over all possible outcomes.

The expected value, which estimates the average number of points per draw, is derived from knowledge of the probability distribution—the p values. If a large number of balls is actually taken from the bag and the points recorded, then the mean value of those points, as given by Rule 4, is estimated by the expected value; the estimate is more accurate the larger the sample size (that is, the more balls removed).

When the occurrence of events is governed by the laws of probability, knowing the expected value is usually the first concern. For example, the annually averaged precipitation at a location is a useful statistic if you want a general sense of how wet or dry the location is. But the variation around the average is also of great interest. Living in a climate with nearly the same precipitation year after year is quite different from enduring very wet years and very dry ones, even if the average over many years is the same in both locations.

A mathematical quantity called the *variance* is often used to describe variation. Consider a long list of numbers, each of which is an event or outcome. They might, for example, be yearly rainfall totals for many years. Denoted by the symbol σ^2, the variance in the data set is a measure of the degree to which the numbers in the list differ from their average. More precisely, it is defined as the average value of the square of the difference between each number and the average of the numbers. If the actual values of the numbers are again labelled n_i and there are N of them, then the variance in the data set is given by Rule 5:

RULE 5:

$$\sigma^2 = \left(\frac{1}{N}\right)\sum_{i=1}^{N}(n_i - \langle n \rangle)^2.$$

Given a probability distribution, the expected value of the variance is

RULE 5′:

expected value of variance

$$= \sum(\textit{outcome} - \textit{average value of outcome})^2(\textit{probability of outcome}).$$

Another frequently encountered measure of variation is the standard deviation of the outcomes, defined simply as the square root of the variance:

$$\text{standard deviation} = \sqrt{\sigma^2}.$$

Finally, we have a useful approximate formula for calculating the variance of a function of some variable when the variance of that variable is known. Suppose that the variance of the variable x is ${\sigma_x}^2$. Let $f(x)$ be some smoothly varying function of x, and let ${\sigma_f}^2$ be the variance of f that results from the variance in x (that is, we are ignoring other contributions to the variance of the function that might result from variation in the parameters that characterize the function). Then,

RULE 5″:

$${\sigma_f}^2 \cong \left(\frac{df}{dx}\right)^2 {\sigma_x}^2.$$

The more curvature there is in the function $f(x)$, the less reliable is this estimate of ${\sigma_f}^2$.

Armed with these rules, we can now tackle Problems I-1 through I-5.

EXERCISE 1: Show that the two bold-faced expressions for the probability of drawing at least one red ball in two draws are identical.

EXERCISE 2: What is the probability that you will not be strike-plagued on a round trip between A and D?

EXERCISE 3: A recent article in the *San Francisco Chronicle* stated that there is approximately a 70% chance of a major earthquake in the San Francisco Bay region by the year 2030. An accompanying map showed the seven major faults in the area and listed the probability of a big earthquake on these faults between now and 2030 as being 32%, 6%, 6%, 4%, 18%, 21%, and 10%. Assuming big earthquakes on these faults are independent events, do you agree with the ~70% estimate for the probability of a "big one" in the Bay Area?

Problems

I-1. Pick Your Poison

Many people assume that the probability of cancer or other health effect from exposure to radiation is directly proportional to the dose. This assumption is called the linear hypothesis (LH), and considerable evidence supports it. Suppose 100 people are each exposed to a dose D, and that as a result 10 people die from the radiation. The dose D thus results in a 10% chance of death for each person. According to the LH, if people are exposed to a dose of $D/5$, we would expect a 2% chance of death for each person; so if 500 people are each exposed to $D/5$, we would again expect 10 of them to die.

Some argue that the LH cannot be valid because it leads to peculiar conclusions when applied to other types of hazard. For example, let's apply it to the risk of death from an aspirin overdose. Suppose, plausibly, that if a person takes 30 aspirin tablets, the chance of death is 50%. According to the LH, if a person takes 1 aspirin, the chance of death is $50\%/30 = 1.67\%$. We know from experience that this value is not correct—if it were true, aspirin would be banned!

The hazard associated with taking aspirin is clearly not well described by the LH; the mechanism by which aspirin affects the body is not consistent with the LH. So let's see if we can discover a mechanism by which pills might hurt us in a manner consistent with the LH. Suppose a pill exists with the property that one out of every 100 such pills contains a lethal amount of a substance, and that the other 99 are harmless. Calculate how many people would be expected to die if 1,000 people each take 1 pill and if 1,000 people each take 10 pills. Does your answer support the LH?

. . . .

The risk of death from a pill is 0.01. If 1,000 people take a single pill, we would expect 10 of them to pick the lethal ones, and so 10 people would die. If the dose is 10 times as great (10 pills per person), what is the chance that a particular person will be lucky and not pick a poisoned one? The first pill carries a 0.99 chance of being safe. Assuming the supply of pills is so huge that removing one pill does not appreciably change the 1% probability that the next pill will be poisoned, the second pill also carries a 0.99 chance of being safe, and so the chances that the person will survive two pills is $(0.99)^2 \sim 0.9801$. If the person takes 10 pills, the chance of surviving is $0.99^{10} \sim 0.9044$. Thus, the chance of not surviving is $1 - 0.9044 = 0.0956$. Hence, out of 1,000 people each taking 10 pills, an average of 95.6 will die. This probability is almost 10 times the number expected to die from taking 1 pill, and so the LH is compatible with this "spiked pill" model.

EXERCISE 1: Calculate the expected number of deaths resulting from 1,000 people each taking 50 such spiked pills, and then for 100 and for 500 pills. Does the LH break down at large doses of spiked pills? Assuming the LH is correct for low doses of radiation, why is it likely to break down at very large radiation doses?

EXERCISE 2: If the LH did apply to aspirin, and 30 people each take 1 aspirin, what is the chance that there would be at least 1 death among the 30?

I-2. Infanticide in China?

There are slightly more baby boys than girls under the age of 1 in China today. That, coupled with the strong desire by Chinese parents to produce a male heir, has been cited by some as evidence that female infanticide is practiced in rural areas of China. In 1983, when I was in China attending a conference, I noticed in a Chinese newspaper an attempt to counter such claims—the paper published a "scientific explanation" for the slightly skewed gender ratio. The reason for the gender imbalance, it said, was that some families will continue having children until they get a boy, and then under pressure from the Chinese government to limit family size, they will stop having more kids. Thus, if the first child is a boy, they will stop there. If the first is a girl and the second is a boy, they will stop at two, and so on. Sounds like a reasonable explanation, right? Could it explain a gender imbalance among young Chinese?

. . . .

Let's suppose that any given pregnancy and birth is as likely to produce a boy as a girl and that the two genders have equal survival rates during childhood. (This set of assumptions is not exactly true, but it is what the "expert" in the newspaper was implicitly assuming.) Suppose, further, that parents continue to bear children until a boy is born, and then they stop having children. What would the gender ratio be under those circumstances?

The best way to approach this problem is to think about all the various possible outcomes for gender assignments within different-sized families, assign a probability to each, and then work out the expected outcome. Because survival is assumed to be independent of gender, we need only think about births. To assign the probabilities, then, we need to recall only two things:

1. The probability that any particular newborn (the second born, say) is a particular gender (a girl, say) is $1/2$.
2. By Rule 2 from the Background section, the joint probability of a sequence of births is the product of the probabilities for each birth in the sequence. This is reasonable, provided parents do not have a natural propensity to produce children of a particular gender; if

you have just had a string of 3 girls, the probability that the next one will be a girl is still the same, according to this assumption, as was the probability that the first of the 3 would be a girl—namely, 1/2.

Thus, we get the following table of outcomes and probabilities:

Outcome	**Probability**
The 1st child is a boy	$\frac{1}{2}$
The 1st is a girl, 2nd a boy	$\frac{1}{2} \times \frac{1}{2} = \frac{1}{4}$
The 1st and 2nd are girls, 3rd is a boy	$\frac{1}{2} \times \frac{1}{2} \times \frac{1}{2} = \frac{1}{8}$
.	.
.	.
.	.

Every family, regardless of which of the outcomes above it finds itself in, has exactly one boy, and so

$$\langle \text{number of boys born to a family} \rangle = 1. \tag{1}$$

Actually, I slipped an implicit assumption in here. To be able to assert that every family has 1 boy, we have to assume that procreation can continue indefinitely until a boy is born, even if that occasionally means a family will have 9,420 girls before the boy is born! We will stick with that preposterous assumption for the remainder of the calculation, but in an exercise, you will get a chance to see what happens under a far more reasonable assumption.

To derive the ratio of births of boys to births of girls in this society, we next have to work out the average number of girls born per family, which we can do from the numbers in the table. Recalling Rule 4 in the Background section, the average number of girls born is equal to the sum of the product of the number of girls in each outcome multiplied by the probability of that outcome, or

$$\langle \text{number of girls born per family} \rangle$$
$$= \sum (\text{number of girls per outcome})(\text{probability of that outcome})$$
$$= 0\left(\frac{1}{2}\right) + 1\left(\frac{1}{4}\right) + 2\left(\frac{1}{8}\right) + 3\left(\frac{1}{16}\right) + \frac{4}{32} + \cdots \tag{2}$$

How do we compute that sum? You can either look it up in the Appendix, where it is of the form $(1/2)\sum_{n=0}^{\infty} n(1/2)^n$, or you can use the following trick:

Denote the sum in Eq. 2 by the symbol S. Then note that

$$\frac{S}{2} = \frac{1}{8} + \frac{2}{16} + \frac{3}{32} + \frac{4}{64} + \cdots = S - \frac{1}{4} - \frac{1}{8} - \frac{1}{16} - \frac{1}{32} \cdots.$$

Now denote by S':

$$S' = \frac{1}{4} + \frac{1}{8} + \frac{1}{16} + \cdots .$$

So,

$$\frac{S}{2} = S - S'.$$

S' can be evaluated by noting that

$$2S' = \frac{1}{2} + \frac{1}{4} + \frac{1}{8} + \cdots = S' + \frac{1}{2}.$$

Hence,

$$S' = \frac{1}{2}.$$

Then,

$$\frac{S}{2} = S - \frac{1}{2},$$

or $S = 1$. In other words, on average, 1 girl is born to each family.

So we have shown that under the stated assumptions, the average number of girls born to each family will equal the average number of boys born to each family. The explanation in the newspaper is false.

EXERCISE 1: Let's avoid the assumption of potentially unlimited procreation (i.e., infinite sums in the above problem) and make a more realistic model. Suppose that procreation continues in each family until either a boy is born or, if no boys are born, 4 girls are born. No family has more than 4 children. What will be the average number of boys and the average number of girls born to each family in that society?

EXERCISE 2: Try to invent a rule governing every family's procreation (e.g., stopping after the second male or the third girl, whichever comes first), such that, if every family obeys it, a skewed gender ratio will result.

I-3. Caesar's Last Breath

How likely is it that at least one nitrogen molecule exhaled by Caesar in his last breath will be in the next breath you take?

• • • • • • • •

Quite slim odds you would guess, right? Well, let's see. Textbook exercises in probability often involve reaching in and grabbing red or blue balls from a bag; here we have to think about grabbing nitrogen molecules from the atmosphere, but it is the same idea. Clearly, solving the problem requires first figuring out what fraction of the nitrogen molecules in the present atmosphere are ones that were in Caesar's last breath. We proceed in easy steps.

First, how many nitrogen molecules were in that last breath? From Appendix XV in *COW-1*, we learn that a typical breath contains about 1 liter of air at standard temperature and pressure (STP). A mole of air at STP occupies 22.4 liters and contains Avogadro's number ($\sim 6 \times 10^{23}$) of molecules.[1] Hence, that last breath contained $1/22.4$ of a mole of air and $6 \times 10^{23}/22.4$ air molecules. Of those molecules, about 78% are nitrogen (Appendix V of *COW-1*), so $(0.78) \times (6 \times 10^{23})/22.4$ nitrogen molecules were expelled to the atmosphere.

The whole atmosphere contains 1.8×10^{20} moles of air (Appendix III of *COW-1*), or $(6 \times 10^{23}) \times (1.8 \times 10^{20})$ molecules of air, and again 78% of those are nitrogen. Hence, the fraction of all the nitrogen molecules in the atmosphere that were exhaled in Caesar's last breath was

$$\frac{\text{number in breath}}{\text{number in atmosphere}} = \frac{(0.78) \times (6 \times 10^{23})/(22.4)}{(0.78) \times (6 \times 10^{23}) \times (1.8 \times 10^{20})}$$
$$= \frac{2.1 \times 10^{22}}{8.42 \times 10^{43}} = 2.5 \times 10^{-22}. \quad (1)$$

By the same reasoning, in your next breath you will inhale $(0.78) \times (6 \times 10^{23})/22.4 = 2.1 \times 10^{22}$ nitrogen molecules.

Now the problem is beginning to look more like the familiar red-and-blue-balls-in-the-bag situation. One concern you might have, however, is whether the nitrogen molecules in your next breath (I hope you haven't been holding your breath up to this point) are a random sample of those last exhaled by Caesar. In other words, have Caesar's nitrogen molecules mixed fairly uniformly within the atmosphere by now or, for example, might they still be concentrated somewhere over Rome? The mixing time for the entire atmosphere is on the

---

1. Both these facts are in Appendix II in *COW-1*, but they crop up frequently in science, so why not memorize them?

order of years, not even centuries, let alone millennia. Thus, within a few years, any nitrogen molecules localized initially will be uniformly mixed within the atmosphere. Indeed, after two millennia, Caesar's nitrogen is well mixed into the atmosphere.

So now we can apply standard probabilistic reasoning to answer the question. This situation is qualitatively similar to one in which you pull, say, 20 red balls out of a huge bag containing a vast number of balls and in which one-fourth of the balls are red—intuitively, you will agree that the probability in that case is high that you will get at least one red ball. More quantitatively, the probability that the *first* ball selected is not red is $1 - 1/4 = 3/4$. The probability that the *second* ball is not red is also $3/4$ (because the bag has so many balls in it, the probability does not change significantly just because one ball has been removed), so the probability that neither ball is red is $(3/4)^2$. The probability that at least one red ball was selected in two draws is (1 – the probability that both balls were blue), or $1 - (3/4)^2$. Continuing in this way, for our bag analogy the probability of drawing at least one red ball will be $1 - (3/4)^{20} = 0.997$.

For our situation, the number of molecules inhaled is 2.1×10^{22}. This portion is analogous to the 20 balls drawn from the bag. The fraction of the nitrogen molecules in the air that were exhaled by Caesar (2.5×10^{-22}; Eq. 1) is analogous to the fraction $1/4$ in the colored-balls example above.

Hence, the probability, P, of breathing at least one of Caesar's nitrogen molecules is given by

$$P = 1 - [1 - (2.5 \times 10^{-22})]^{2.1 \times 10^{22}} = 0.995. \tag{2}$$

So your next breath almost certainly does contain at least one of Julius Caesar's last exhaled nitrogen molecules!

"N_2, Brute?"

You are probably curious about how I evaluated the expression above numerically. If you tried to do it on a pocket calculator, you probably got an answer of zero (because of a problem with significant figures). Here is the trick I used. In the limit $r \to 0$, the expression $(1-r)^{1/r} \to e^{-1}$, where e is the base of the natural logarithm system (~ 2.718). Hence, letting $r = 2.5 \times 10^{-22}$, and noting that the exponent 2.1×10^{22} can then be rewritten as $5.25/r$, our expression can be rewritten as $1 - (1-r)^{5.25/r} = 1 - [(1-r)^{1/r}]^{5.25} = 1 - e^{-5.25} = 0.995$.

EXERCISE 1: One possible difficulty with this calculation may still remain. Each year, some nitrogen is removed from the atmosphere by a process called nitrogen fixation. Nitrogen fixation converts nitrogen in the atmosphere to a form used by plants as fertilizer—ammonium or nitrate. After nitrogen enters the plant and the plant dies, the nitrogen will then reside in the soil or in water, be taken up by another plant, or be returned to the atmosphere. The passage of nitrogen among these locations is called the nitrogen cycle. If a sizable portion of the nitrogen in Caesar's last breath now resides in soil, water, or plants in the form of fixed nitrogen, then we may have overestimated the chances of breathing some of it in. Given that there are $\sim 4 \times 10^{18}$ kg (nitrogen) in the atmosphere and the global rate of nitrogen fixation has averaged about 2×10^{11} kg (nitrogen)/year over the past several millennia (Appendix XIII in *COW-1*), explain why the correction made for fixed nitrogen is not going to affect our answer very much. Can you think of any other possible corrections to our calculation?

EXERCISE 2: How likely is it that at least one of the hydrogen atoms in Julius Caesar's last drink of water is in the next glass of water you drink? State your assumptions.

EXERCISE 3: We have seen that the probability is quite high that at least one of Caesar's exhaled nitrogen molecules is inhaled by you in your next breath. What is the average number of such molecules you can expect to inhale in your next breath?

I-4. A Bolt from the Blue?

Following the TWA passenger plane disaster in the summer of 1996 over Long Island Sound, speculation was made that the plane may have been hit by a meteor. Is that idea absurd? (For this problem, you need two pieces of information that your own powers of estimation and observation will not provide: on the order of 3×10^3 meteors penetrate to the lower atmosphere each day with enough energy to destroy an airplane if they hit a critical component of the plane, and very roughly about 5% of the area of an airplane, viewed from the top, is critical in that sense.)

. . . .

First, let's work out the probability that at least one plane will be critically struck in a given year. Then we will estimate the probability that the particular TWA plane was struck on that particular flight. Finally, we will discuss which of these two probabilities conveys the best sense of whether the speculation is absurd.

How many planes are in the air at any moment? At commercial airports, planes seem to take off on average about every 5 minutes during the daylight hours (at Chicago's O'Hare and other similiarly busy airports, the rate is actually about one per minute, but at most airports around the world, the rate of takeoff is lower). The number of commercial airports in the world is on the order of 10^3 (100 nations, each on average with 10 major airports; the United States probably has a quarter or more of the total). Thus, during the 12 active hours, or $12 \times 60 = 720$ minutes, of the day, the number of take-offs is approximately $(1000)(720/5) = 144{,}000$. If the typical flight is in the air for 3 hours (a very rough guess), then at any average moment of the 24-hour day, there are on average $(3/24) \times (144{,}000)$ commercial passenger planes in the air, or 18,000 planes.

What is the chance that one of those planes in the air at any moment will be hit by an incoming meteor? The area of an airplane exposed to an incoming meteor is, to round-number accuracy, roughly $40 \text{ m} \times 15 \text{ m}$, or 600 m^2. Because the area of the Earth is $5 \times 10^{14} \text{ m}^2$, each plane "covers" $\sim 10^{-12}$ of the sky. Hence, the 18,000 planes cover 1.8×10^{-8} of the sky. So, when a given meteor arrives, the probability it will hit a plane is 1.8×10^{-8}. If 3×10^3 killer meteors strike each day, the probability that in one day at least one plane will be hit is $3 \times 10^3 \times 1.8 \times 10^{-8} = 5.4 \times 10^{-5}$. Of those hits, about 5% will cause a crash, so the probability of a crash is 2.7×10^{-6} per day. To an appropriate round-number level of precision, this probability is about

10^{-3}/year. So in 50 years of passenger flight at today's volume of air traffic, the chances of a meteor hit causing at least one crash are $\sim$5%.

We also care, though, about the probability that a particular plane would be struck on that particular flight. Let's assume that the effective fraction of area covered by the plane is about twice the 10^{-12} figure we estimated earlier, because 747s and many other transoceanic passenger planes are much bigger than the average commercial passenger airplane. On a flight from New York City to Paris, the plane is in the air for about 6 hours, and thus $(6/24) \times 3 \times 10^3$ meteors strike the earth. The probability that one will hit the plane in a critical spot is then $(6/24) \times 3 \times 10^3 \times 0.05 \times (2 \times 10^{-12}) \sim 10^{-10}$.

One of these probabilities is large enough to be worrisome (10^{-3}), whereas the other (10^{-10}) is nearly negligible. Which one best tells us whether the speculation was absurd? This is not a simple question. Note that if I had asked for the chances of the particular TWA being struck at the *particular place* on its flight trajectory, the probability would have been smaller still. By getting too particular, you can make chances of a rare event seem even rarer! The 10^{-10} figure is of use if you want to estimate your risk when you board a plane for a particular flight; as you can see, you need not be concerned. The 10^{-3} figure yields a sense of whether or not you should be worried about a strike to some plane at some time on some flight. If you were a government investigator of air crashes you would want to know that number. The 10^{-3} figure tells us that after several decades of flight, we should not be too surprised if there is a critical meteor hit. Thus, the speculation is not absurd.

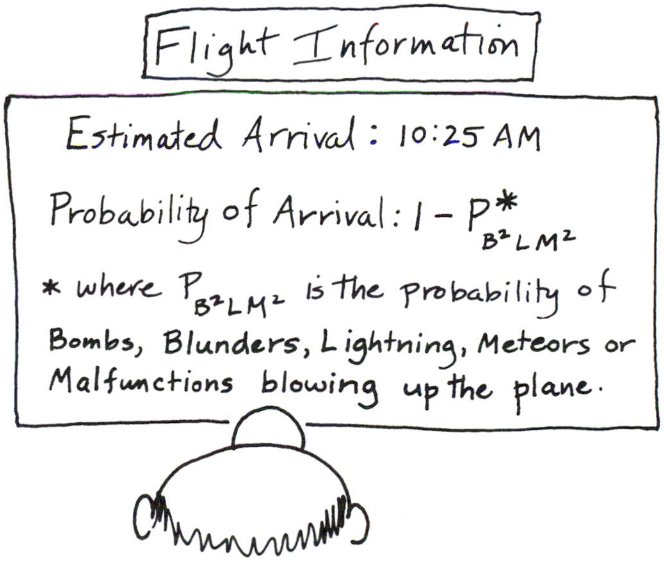

A more relevant question is whether you should be worried about a meteor strike relative to other causes of airplane crashes. The answer is clearly *no*. During far fewer than 10^3 years of commercial passenger plane flight, there have been far more than one crash. We do not expect many crashes from meteor impacts.

When comparing the probability of a meteor hit on a particular flight to the probability of other causes of crashes, you need to be consistent. You can use either answer as long as you proceed in the same way for all causes that you are comparing. Thus, the 10^{-10} result should be compared with the probability of, say, a bomb exploding on a particular flight, and not with the frequency with which bombs destroy airplanes each year. As an aside, if you are comparing dangers and you know something particular about nonmeteoritic dangers in flying—perhaps something about TWA planes, about flights from New York to Paris, about dangers that lurk over Long Island Sound—then you must make the comparison with reference to (that is, incorporating) such particularities.

EXERCISE 1: Lightning strikes the Earth's surface somewhere about once every minute. Assuming that the strikes occur at random locations, that it is lethal to be within 5 meters of a strike, and that you take no defensive action when a storm is nearby, estimate the probability that you will die by being hit by lightning.

I-5. On the Street Where You Live

On the street where you live are 120 residents. Last year, 6 of them received a diagnosis of cancer. Should you move?

. . . .

A cancer cluster is an occurrence of cancer in several people living or working near one another. When such a cluster is observed, we naturally wonder whether there is an environmental cause—perhaps a hazardous chemical waste dump had been located in the neighborhood and houses later built on top of it, or factory workers were exposed to a potentially toxic chemical. This problem is about estimating how concerned we should be and whether we should seek an unusual, localized environmental cause when a cluster is observed.

Estimating the abnormality of the observed cluster begins by using the same type of reasoning used to estimate the probability that a repeated coin toss will yield 6 heads in a row. Because the probability of heads on any given toss is one-half, and repeated tosses are independent of one another, the probability of tossing 6 heads in a row is $(1/2)^6$, or 0.0156. The smallness of that number does not necessarily mean we should doubt the fairness of the coin if we see 6 heads in a row; it all depends on context. Suppose, for example, that each of 100 people flip a coin 6 times. The probability is high that at least one of them will toss 6 heads. Yet if you happened to witness one of those "lucky" persons and you did not know about the 99 other coin tossers, you would be tempted to conclude that the coin used was unfair. In fact, it would be your location, not the coin, that was unusual!

Similarly, with cancer, we would like to know not only how unusual it is for your particular block to have a cancer cluster but how many such clusters occur anywhere each year for reasons that have nothing to do with environmental contamination or even geography. If the chance of *your block's* being the site of a cluster is tiny but the chance of many such clusters occurring *anywhere* each year is high, then you might think differently about the implications of seeing a cluster on your block. In the second half of our calculation, I will address that issue for the cancer occurrence.

We will do the calculations for the United States, but the concept behind the calculation would be useful for any geographic area. In the United States, approximately 10^6 cases of cancer are diagnosed each year, out of a population of 2.6×10^8. The probability that in any given year a particular person on your block is diagnosed with cancer, then, is $10^6/(2.6 \times 10^8) = 0.0038$.

Suppose two people are singled out—what is the probability that in both a cancer is diagnosed? Assuming the probabilities of cancer in each of them are independent of one another (an assumption that would be violated if there was an unusual environmental cause of cancer in the neighborhood), then the probability of both having a cancer diagnosed in a particular year is $(0.0038)^2$. For 6 particular households to have a single cancer each, the probability is $(0.0038)^6 = 3.01 \times 10^{-15}$, which is small indeed.

Note that this small number is actually the probability that 6 particular people will have cancer diagnosed in one year, irrespective of what is true about cancer in the other residents of the block. Because we are interested for now in the probability that 6 and only 6 people would receive diagnoses of cancer, we might want to multiply the probability above by the probability that the other $120 - 6 = 114$ neighborhood residents do not receive cancer diagnoses. As you will see when you do Exercise 3, the latter probability is 0.65, and so the probability that 6 and only 6 particular people receive cancer diagnoses is $0.65 \times 3.01 \times 10^{-15} = 2.0 \times 10^{-15}$.

We are not finished, however, for the problem statement was that 6 people out of 120 on the block had a cancer diagnosed. It didn't state which people, yet we derived the probability that 6 *particular* people had cancer diagnosed. To calculate the probability that *some* group of 6 people out of 120 were afflicted, we have to multiply the probability for a particular set of 6 persons by the number of ways that any 6 can be chosen out of 120.

The number of ways to choose n items out of a larger list of N items is frequently encountered in probabilistic analyses. Mathematicians have given it a special name and symbol. The name is "N, choose n," and the symbol is $\binom{N}{n}$.

Our task here is to calculate "120, choose 6," or $\binom{120}{6}$. The first of the 6 can be any of the 120. The second can be any of the remaining 119, the third any of 118, and so on. Thus, there are (120)(119)(118)(117)(116)(115) ways to pick the 6. This calculation, however, isn't quite right, because it counts as separate the following permutations, which are identical pairs:

1. The first cancer is in person number 5 and the second is in person number 18
2. The first cancer is in person number 18 and the second is in person number 5.

Although those permutations might be considered different if we were keeping track of the order and types of cancers, we are not given such information, and our only concern is with the combinations of 6 incidents among 120 locations. So to correct for this overcounting, we have to divide by the number of ways to arrange 6 objects in different orders, or permutations. The first can be any of the 6, the second any of the remaining 5, and so on, and thus there are (6)(5)(4)(3)(2)(1) essentially equivalent ways in which the 6 cases can be arranged within any 6 of the 120 residents. Hence,

$$\binom{120}{6} = \frac{120!}{[(6!)(120-6)!]}$$
$$= \frac{[(120)(119)(118)(117)(116)(115)]}{[(1)(2)(3)(4)(5)(6)]}$$
$$\sim 3.65 \times 10^9, \qquad (1)$$

where we have used the symbol ! to indicate a factorial: $n! = n(n-1)(n-2)(n-3)\ldots(2)(1)$. Multiplying this number of combinations by the probability of any given combination occurring, 2×10^{-15} yields the probability of the particular block having 6 cancer cases diagnosed in one year. Calling this probability P_6, where the subscript indicates that 6 persons had a cancer case diagnosed last year, we get $P_6 = 7.3 \times 10^{-6}$.

That probability is small, so if the situation actually did arise on your block, your initial reaction would probably be one of concern. But how should we interpret the situation? Suppose you lived on a block with no other residents, and you had cancer diagnosed last year. Would you suspect something unusual? Most people would be greatly upset by the event but not suspicious that there might be some unusual cause. In that case, the probability would be 0.0038, as estimated above. How much lower than this does the probability have to be to warrant our suspecting something unusual? There is no unambiguous answer to this—it really is up to you in the last analysis.

We can, however, delve further into the situation with a little more of the same type of mathematics. As mentioned above, with enough observers of strings of coin tosses, at least some are likely to witness the tossing of 6 heads in a row. In the case of neighborhoods with cancer clusters, we thus might want to estimate how many such neighborhoods with cancer clusters there should be in the United States in an average year if the cancers are randomly occurring. If the answer comes out that, say, 20 such clusters should be observed somewhere in the United States each year, then we would want to know how many were actually observed. If the answer was considerably higher than 20, then we would suspect that cancers are unusually clustered and that some local (and therefore probably environmental) cause is likely for the excess above the number 20 expected in the random model.

Let's make a very simplified model of the neighborhood structure of the United States. We assume all people live on blocks with 120 persons. The difference between this simple assumption and the real world is big, but in fact it will not matter to our overall approximate finding. With 120 persons per block, there are a total of $2.6 \times 10^8/120 = 2.2 \times 10^6$ blocks. We will denote that number by n_{max}, the maximum number of blocks in which 6 cancers might possibly have been diagnosed last year. The probability of a particular one having a cluster of 6 cancers is $P_6 = 7.3 \times 10^{-6}$. What is the expected number of blocks with such clusters?[2]

We denote by $P_6(n)$ the probability that 6 people on each of n and only n blocks had a cancer case diagnosed last year. Calculating the value of $P_6(1)$ is similar to the problem of calculating the probability of, say, 5 coin tosses yielding one and only one heads. That probability is given by (the probability of any one particular coin toss being heads) × (the number of tosses) × (the probability that the other 4 tosses are not heads). The probability of any particular toss being heads is of course 1/2. This is the number analogous to our quantity $P_6 = 7.3 \times 10^{-6}$. The number of tosses is 5, analogous to our quantity $n_{max} = 2.2 \times 10^6$. The probability that the other 4 tosses are not heads is just $(1 - 1/2)^{5-1} = 1/16$; in our case, the analogous quantity is $(1 - P_6)^{n_{max-1}}$.

For the coin toss situation, the answer then is $P(1) = 5(1/2)(1/2)^4 = 5/32$. In similar fashion, for our situation,

$$\begin{aligned} P_6(1) &= n_{max}(P_6)(1 - P_6)^{n_{max}-1} \\ &= (2.2 \times 10^6)(7.3 \times 10^{-6})(1 - 7.3 \times 10^{-6})^{2.2\times10\wedge6-1}. \end{aligned} \tag{2}$$

The third term on the right side of this expression is evaluated by the same trick we used in the Caesar's breath problem in this section, and is nearly equal to $e^{-16.0} = 1.1 \times 10^{-7}$. Hence,

$$P_6(1) \sim 1.8 \times 10^{-6}. \tag{3}$$

What about $P_6(2)$? It is given by (the probability of two specific blocks having clusters) × (the number of ways to choose 2 blocks out of a total of n_{max}) × (the probability that the other $n_{max} - 2$ blocks do not have clusters). Assuming that cancer incidences in different blocks are independent events, the first of these three terms is just the square of P_6. The second is given by the factorial expression for "n_{max}, choose 2." The third is given by $(1 - P_6)^{n_{max-2}}$. Hence,

$$\begin{aligned} P_6(2) &= [(P_6)^2]\left[\frac{n_{max}!}{2!(n_{max} - 2)!}\right][(1 - P_6)^{n_{max}-2}] \\ &\sim 1.4 \times 10^{-5}. \end{aligned} \tag{4}$$

---

2. In actuality, we might want to accept as a rare event a diagnosis of 6 *or more* cancers per neighborhood and ask how many of those are expected each year. This is left as an Exercise (9) for the reader.

Continuing in this way, the following general formula results:

$$P_6(n) = [(P_6)^n]\left[\frac{n_{\max}!}{\{n!(n_{\max} - n)!\}}\right][(1 - P_6)^{n_{\max} - n}]. \quad (5)$$

We have just derived a very famous probability distribution—the binomial distribution! It arises over and over again in statistics. As we saw in the Background section, if we have any probability distribution, $P(n)$, for a variable n (where P is not necessarily a binomial distribution), then the expected value of the variable n described by that distribution is given by

$$\text{expected value of } n = \sum_{n=0}^{n_{\max}} nP(n). \quad (6)$$

This equation can be evaluated for the binomial and it yields a wonderfully simple and exact answer (see Exercise 8):

$$\text{expected value of } n = n_{\max}P_6. \quad (7)$$

For our case, this equation is $(7.3 \times 10^{-6})(2.2 \times 10^6) = 16$.

Thus, we would expect that each year, in an average of 16 blocks around the United States, 6 cancer cases will be diagnosed. The model that we deployed to derive this result is sometimes called a "random model." This means that we assumed that the large pool of cancers occurring each year was randomly distributed around the United States—there were no hot spots.

How does our result help you think about what happened on your block? It is actually of less help to you than it is to the Surgeon General or other health officials concerned about the health of the United States as a whole. At the scale of the United States, if, say, 100 such clusters were found in a particular year, then the pressure would mount to search for carcinogens in the neighborhoods with clusters. Moreover, if only 16 clusters were observed in the entire United States next year, but if they were all concentrated in, to pick a random state, New Jersey, then we would be far more concerned than if they were widely distributed among all the states.

If well over 16 clusters (100) are observed in the United States next year, and they are widely distributed, statistical analysis, by itself, cannot help you decide whether the cluster on *your* block was one of the 16 expected clusters or one of the excess that might well be the result of some unusual local environmental problem. Your neighborhood might be perfectly safe (one of the expected 16) but surely we should be suspicious about *most* of the clusters because 100 is so much greater than 16.

But is it really? How much uncertainty is there in that figure of 16?

There are many sources of uncertainty in our result. Among these are our simplified assumption about neighborhood composition and imperfect data on total cancer incidence each year in the United States. The kind we will look at here is different, however—the uncertainty arising simply because we are dealing with a probabilistic phenomenon. As we saw in the Background section, statisticians describe such uncertainty with a quantity called the standard deviation. For the binomial distribution, the expected value of the standard deviation has been worked out by mathematicians:

$$\text{expected value of standard deviation of } n = [n_{\text{max}} P_6 (1 - P_6)]^{1/2}. \quad (8)$$

Evaluating this equation, we get a standard deviation of $\sim\sqrt{16} = 4$. This result should be interpreted in the following way. The probability that an observed value of n lies within a standard deviation of the mean, and thus between $16 - 4 = 12$ and $16 + 4 = 20$, is 68.27%.

The probability that a value of, say, 100 would be observed can be worked out as well. To do this, we could evaluate the binomial expression in Eq. 5 by brute force, but let's be more clever—it will teach us something new. To simplify the notation, we rewrite the binomial distribution (Eq. 5) in the slightly less cumbersome form:

$$P(x; N, p) = p^x \left[\frac{N!}{x!(N-x)!}\right] (1-p)^{N-x}. \quad (9)$$

Here, N is the number of "throws" or blocks, p is the probability of an event occurring in a given throw or a particular block, and x is the variable whose probability we seek: an event showing up on x throws or x blocks. We define a new variable, λ, by $\lambda = Np$, and from here on, we will assume that N is very large compared with x, and p is very small (equivalently, λ/N is very small). Recall that in our cancer problem, N was a large number, p a small one, and the product, λ, was neither much bigger or much smaller than 1. In terms of our new variable, the binomial distribution can be rewritten:

$$P(x; N, \lambda) = (\lambda/N)^x \left[\frac{N!}{x!(N-x)!}\right] (1 - \lambda/N)^{N-x}. \quad (10)$$

Using our old trick $(1-r)^{1/r} \to 1/e$ as $r \to 0$, the second term on the right side becomes $e^{-\lambda}(1-\lambda/N)^{-x}$, and for λ/N small, this is $\sim e^{-\lambda}$. The third term, "N choose x", can be rewritten as $N^x(1-1/N)(1-2/N)\ldots(1-x/N)/x!$. In the limit of N being large compared with x, this is $\sim N^x/x!$. Thus, combining terms, we get

$$P(x; N, \lambda) \sim \left(\frac{\lambda}{N}\right)^x \frac{e^{-\lambda} N^x}{x!}. \quad (11)$$

After cancelling a factor of N^x, this approximation becomes

$$P(x; N, \lambda) \sim \frac{\lambda^x e^{-\lambda}}{x!}. \tag{12}$$

This is another famous distribution, the Poisson distribution. Note that it is independent of N. As can be seen from the derivation, it is a limiting case of the binomial distribution. It is valid when (a) N (the number of tosses, or the number of blocks, or more generally, the number of possible situations in which the x events could arise) is very large compared with x, and (b) $p = \lambda/N$ (the probability of 6 heads in a row, or a block having a cancer cluster, or more generally, some particular event of interest occurring) is very small. Then it tells us the probability of x events of interest occurring. The Poisson distribution has the property that in the limit in which x approaches its mean value, λ, the distribution approaches the normal (or Gaussian) distribution (Exercise 10).

Now we can return to the query that started us down this path: what is the probability of 100 clusters occurring? To evaluate this question, we let x (or n in our old notation) $= 100$, $\lambda = pN$ (or in our old notation, $P_6 n_{\max}$) $= 16$, and get

$$P_6(100) = \frac{e^{-16} 16^{100}}{100!}. \tag{13}$$

Although this equation may not seem like a very simple expression, evaluating it is far easier than evaluating Eq. 5. One further simplification will help us however: Stirling's formula. This handy approximation states that for large x,

$$x! \sim (2\pi x)^{1/2} x^x e^{-x}. \tag{14}$$

Hence,

$$100! \sim 628^{1/2}100^{100}e^{-100} \tag{15}$$

and thus

$$\begin{aligned} P_6(100) &\sim e^{-16}16^{100}e^{100}100^{-100}(628)^{-1/2} \\ &= \left(\frac{16e}{100}\right)^{100} e^{-16}628^{-1/2} \\ &= 0.435^{100}e^{-16}628^{-1/2} \\ &= 3.2 \times 10^{-45}. \end{aligned} \tag{16}$$

The chances of 100 such clusters occurring are small indeed!

We started this problem with a description of a very personal issue —a cancer cluster on your block—and posed a question of immediate concern to you. The mathematics, however, led us to a more impersonal conclusion. It did not shed much light on the plight of any particular individual or block, but rather it provided a rigorous basis for a more aggregated policy decision at the level of a large number of blocks and a possibly large number of cancer clusters. This situation arises frequently in statistical analyses—it is intrinsic to such analysis that we are forced to conclusions at a level of aggregation appropriate to the nature of the data. In this case, we could not squeeze out of a single observation of a cluster any sound conclusion about the likelihood of a localized environmental cause for that cluster, but a comparison of the expected number of such clusters (~ 16) and the actual number could lead to an important policy decision by public health researchers and policymakers. It must be understood, however, that in the United States today, we unfortunately lack good data about the incidence of cancer clusters; in the example above, I just assumed 100 clusters were observed for the sake of argument.

As with nearly any problem concerning human risk, this problem needs an attached warning label. Because we dealt here only with probabilities, the insights we generated can be misleading if misinterpreted. Even a single cancer can indicate an environmental hazard. Calculations like this one cannot really tell you whether or not you should move away from the site of a cancer cluster. They can only give you a sense of the extraordinariness of the event. Personal decisions should be based on much more information than the sort provided here.

EXERCISE 1: We did not say anything about the distribution of the cancer cases among the households comprising the block. If 6 cancers occurred in a total of 2 households, what additional possible explanation, besides an environmental contaminant, would you be tempted to consider?

EXERCISE 2: Suppose a cancer cluster with 6 cancers occurred on your block last year, and a total of 100 such cancer clusters around the United States were observed that year. What is the probability that yours was one of the 16 expected in the random model and therefore not the result of some locally concentrated environmental carcinogen?

EXERCISE 3: Calculate the probability that 114 particular people on the block are not going to receive a cancer diagnosis this year.

EXERCISE 4: Show that the Poisson formula gives nearly the same answers for $P_6(1)$ and $P_6(2)$ that we calculated in Eqs. 3 and 4.

EXERCISE 5: Estimate the value of $P_6(n)$ for $n = 20$, 24, 25, 26, 27, and 30 using the Poisson distribution with Stirling's approximation. From your answers and from the previously computed values for $n = 1, 2$, and 100, what can you conclude about the shape of the Poisson distribution; is it symmetric about its mean?

EXERCISE 6: What is the probability that there were no cancers diagnosed on your block last year? Calculate $P_6(0)$, the probability that no cancer clusters (with 6 cancers) occur in a given year in the United States. [*Hint*: $\binom{n}{0} = 1$]

EXERCISE 7: We carried out our analysis using from the outset a cancer rate—the number of diagnosed cancers in the United States in any particular recent year. Thus, all the above calculations were for a single year. However, a sense of urgency about a local environmental cause of cancer could arise from the observation of a cancer cluster in any year that you live on your block. Discuss qualitatively how the element of time might be incorporated into the analysis.

EXERCISE 8: Starting with the binomial distribution (Eq. 5) rewritten in the form

$$P(n) = p^n(1-p)^{N-n}\frac{N!}{(n!(N-n)!)},$$

prove that the mean value of n, given by

$$\langle n \rangle = \sum_{n=0}^{N} nP(n),$$

is given by pN. [*Hint*: Make clever use of the fact that

$$\sum_{n=0}^{N} P(n) = 1.]$$

EXERCISE 9: In actuality, we might want to accept as a rare event a diagnosis of 6 *or more* cancers per neighborhood and ask how many of those are expected each year. We can define a quantity $P_{\geq 6}$ in analogy with P_6; it is the probability that on a particular block with 120 residents, cancer will be diagnosed in 6 or more in a particular year. Calculate the value of $P_{\geq 6}$ and the average number of blocks in which such clusters will occur. [*Hint*: You will have to calculate the sum from $x = 6$ to 120 over the binomial distribution $P(x\text{: } 120, 0.0038)$. Make use of the fact that the sum from $x = 0$ to 120 is 1, and be very careful with significant figures.]

EXERCISE 10: Show that the Poisson distribution, in the limit that x is close to its mean value of λ, approaches the normal (or Gaussian) distribution: $P(x) \propto \exp[-(x - \lambda)^2/(2\sigma^2)]$.

Chapter II
Optimization

Gonzalo: "Here is everything advantageous to life."
Antonio: "True, save means to live."

Introduction

Optimization is all about achieving the most or the least—the least waste, the fastest process, the shortest route, the biggest return on investment, the maximum profit, the minimum likelihood of causing extinction, the most offspring, the largest biomass.

The most important point to keep in mind about optimization is that the frame of reference—the boundary of the analysis—always excludes issues and factors of concern to some if not all people. Thus, only some aspects, but not all, are optimized in any application of the technique. Optimizers, like Gonzalo, often fail to tell you that point, especially if among the aspects they are not optimizing is the one they suspect you most care about! Some will not tell you this because they want you to accept their objective, and they hope the rigor of their calculation will sugarcoat the pill of their unconvincing objective. Others may truly have lost sight of the multiplicity of human wants.

Lest you think optimizers are lazy for not optimizing everything, however, please note that in general you cannot optimize for more than one aspect at a time. For example, imagine trying to spend your previous month's income in such a manner that you travel the greatest distance on airplanes this month and simultaneously put the most savings aside for your child's education. Somewhat more subtly, imagine finding a place to live that is the closest to where you work and the farthest from a smelly factory. Without more information about the relative value you place on each of the two objectives, you cannot meet both criteria (unless the smelly factory is at the antipode to your workplace!). As we learned in the Introduction to Chapter I, first come values and then comes mathematics; much as one may enjoy formulas and derivations, they will never eliminate the need for people to decide which values should govern choices.

Background

The easy part of optimization is carrying out the mathematical operation itself. This widely used technique, based on differential calculus, is readily grasped from Figure II-1. At both the maximum of the function (point a) and the minimum (point b), the curve is flat, of course, and so the slope of the curve is zero at these points. From differential calculus, the slope of a function, $f(x)$, is given by the first derivative: $df(x)/dx$. Setting the first derivative equal to zero and solving for x gives the value(s) of x where the function is either a maximum or minimum.

To determine which, note that in the neighborhood of a maximum, the slope trends from positive (before the maximum) to zero at the maximum, and then to a negative value. Thus, the slope is decreasing as x increases through the maximum, or $d(\text{slope})/dx < 0$. Conversely, at a minimum, the slope is increasing, or $d(\text{slope})/dx > 0$. Recall, however, that the derivative of the slope is the second derivative of the function itself, so

$$\frac{df(x)}{dx} = 0 \quad \text{at maximum or minimum;}$$

$$\frac{d^2f(x)}{dx^2} < 0 \quad \text{at maximum;}$$

$$\frac{d^2f(x)}{dx^2} > 0 \quad \text{at minimum.}$$

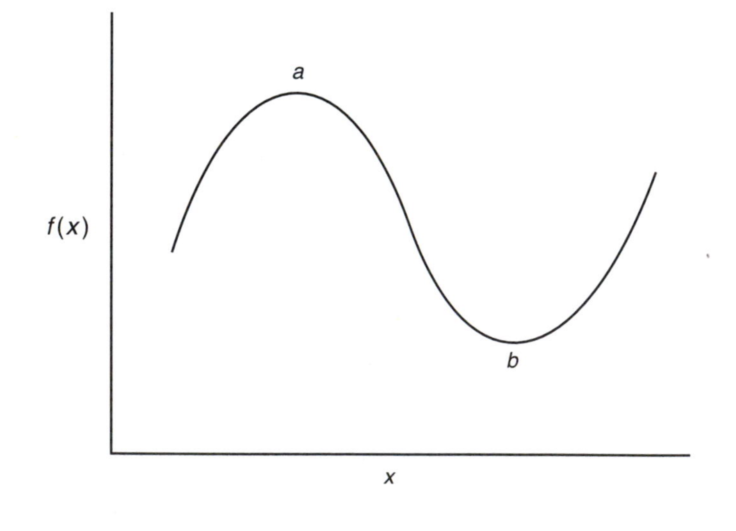

Figure II-1 A curve with a maximum at point a and a minimum at point b.

But what is f, and what is x? That is the hard part of optimization—figuring out what function (f) describes the quantity to be optimized and what parameter (x) we can manipulate until an optimum state is achieved. You will get practice making those choices in the examples that follow.

Often, in real life, optimization has to be carried out in the face of either predictably or unpredictably varying conditions. For example, weather patterns affect agricultural productivity, so farmers seeking an optimum mix of crops need to take both the predictable and the unpredictable variability of weather into account. Examples abound of optimization calculations carried out by policymakers under the unrealistic assumption that the constraints on the variable to be optimized were time-independent, even though the results of such calculations can be highly misleading. Later in the chapter you will have a chance to explore some of the consequences of time dependence (Exercise 4 of Problem II-1 and Exercise 2 of Problem II-3).

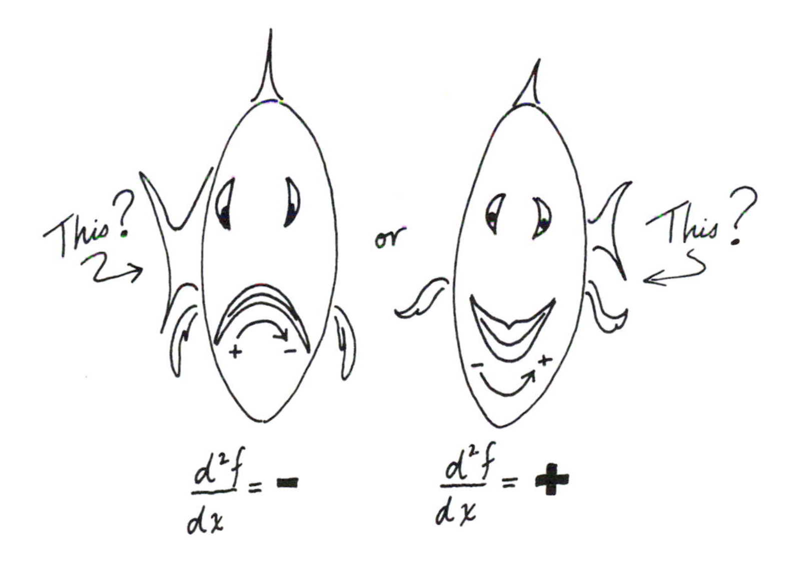

Problems

II-1. Advice for Farmers

A nation has 10^5 hectares of agricultural land to devote to growing some mix of potatoes and tulips. The annual potato yield is 800 kg/ha, the average cost to farmers to grow and harvest potatoes is \$0.05/kg, and they can be sold to a distributor for \$0.08/kg. The annual tulip yield is 4,000 stems/ha and the cost to grow and harvest them is \$0.002/stem. The amount of money farmers can get per tulip declines, however, as the total tulip harvest of the nation increases because demand for tulips is more limited than demand for potatoes and thus the market for tulips can be saturated. The following is a reasonable approximate formula for the sale price of tulips, P, which incorporates the notion that the tulip market can be saturated: $P = pN_0/(N_0 + N)$. In this expression, P is the average price per tulip stem, p is \$0.01/stem, N is the number of harvested tulip stems, and $N_0 = 10^6$ stems. If the farmers in the nation get together to decide on the best strategy to maximize the total national income from the 10^5 ha, what fraction of the nation's land will they devote to tulips?

. . . .

As with many problems involving optimization, a critical first step is defining variables carefully and writing the variable to be optimized as a function of the variable over which you have some control. In this case, total net income is the variable to be maximized, and the fraction of land to be planted in tulips is the "control variable." Let Y be the total net income from the 10^5 hectares. We also define f to be the fraction of the land devoted to tulips, so that $1 - f$ is the fraction of the land on which potatoes are grown. Note that the number of harvested tulips, N, can be expressed in terms of f: $N = 4{,}000 \times f \times 10^5$.

The tulip yield is $4{,}000 \times f \times 10^5$ stems/year.

The potato yield is $800 \times (1 - f) \times 10^5$ kg/year.

The cost of growing the tulips is $0.002 \times 4{,}000 \times f \times 10^5$

$= 0.8f \times 10^6$ \$/year.

The cost of growing the potatoes is $0.05 \times 800 \times (1 - f) \times 10^5$

$= 4(1 - f) \times 10^6$ \$/year.

The gross income from the tulips is $PN = \dfrac{pN_0N}{N_0 + N}$

$$= \frac{0.01 \times 10^6}{10^6 + 4{,}000 \times f \times 10^5} \times 4{,}000 \times f \times 10^5$$

$$= \frac{4f}{1 + 400f} \times 10^6 \text{ \$/year.}$$

The gross income from the potatoes is $0.08 \times 800(1 - f) \times 10^5$

$= 6.4(1 - f) \times 10^6$ \$/year.

The net income, Y, is the gross income minus costs. Expressing all our results in units of \$$10^6$, we have

$$Y = [\text{tulip net income}] + [\text{potato net income}]$$
$$= \left[\frac{4f}{1 + 400f} - 0.8f\right] + [6.4(1 - f) - 4(1 - f)]$$
$$= \frac{4f}{1 + 400f} - 3.2f + 2.4.$$

To find the extremum (minimum or maximum) of this function $Y(f)$, we have to set the derivative dY/df equal to zero and solve for f. The derivative is easily evaluated:

$$\frac{dY}{df} = \frac{(1 + 400f)(4) - 4f(400)}{(1 + 400f)^2} - 3.2 = \frac{4}{(1 + 400f)^2} - 3.2.$$

Hence our task has become to find the root of the equation $4/(1 + 400f)^2 = 3.2$. Rearranging this equation, we get $1 + 400f = (4/3.2)^{1/2} = \pm 1.118$. This type of result often happens in mathematics —two or more answers are possible, where a unique one was anticipated. As is also often the case, however, a good reason exists to show why only one of the answers is really possible. If the -1.118 value is chosen, then f turns out to be negative. That's impossible, so we should look at the positive answer. Choosing the positive sign, $400f = 0.118$, or $f = 0.000295$.

Is this solution a minimum or a maximum net income strategy? To determine this, we take the second derivative d^2Y/df^2. This expression is equal to $d(1+400f)^{-2}/df = -2(400)(1+400f)^{3}$. Substituting in the value $f = 0.000295$, we see that this expression is negative, and hence the solution we obtained is a maximum income solution. The nation should put only 29.5 hectares into tulip production.

EXERCISE 1: Show that the formula $P = pN_0/(N_0 + N)$ does indeed incorporate the notion that the tulip market is saturable.

EXERCISE 2: What total average net income per hectare will the nation derive from this optimal strategy?

EXERCISE 3: Who makes more money per hectare if the optimum strategy is adopted—the potato farmers or the tulip farmers? Would it pay for a potato farmer to cheat and surreptitiously grow tulips on a hectare or two of his land?

EXERCISE 4: How would you build a risk factor into this analysis? More specifically, suppose the risk of growing tulips is 0.05; in other words, there is a 5% chance in any given year that the tulip harvest will fail because of climate. The risk of a potato crop failure is 0.02. If the goal is to maximize expected net income, what should the mix of tulips and potatoes be?

II-2. Efficient Censusing

This problem will refresh your memory about probability as well as give you more practice in optimization.

> The mark-release-recapture technique for censusing a population of, say, fish in a lake is widely used by biologists. It works as follows. You catch some subsample of the fish population, mark the fish somehow so that the mark will persist, and return the marked individuals to the wild. You do not want the mark to bother the animal too much, of course, so for fish you might take a tiny, distinguishable clipping from a fin. You then come back a little later and catch another subsample. By counting the fraction of the second catch that are marked you can estimate the population size. Assume that you have a fixed total amount of time to devote to fishing and that the number of fish you catch is proportional to the time you spend fishing. You therefore can adjust the relative sizes of the first and second catches by the way you allot your total fishing time between the two catches. The question is: How should you divide your time between the first catch and the second catch so as to minimize the error in your estimate of the size of the fish population?

. . . .

Let N_0 be the actual population of fish in the lake, let N_1 be the number of fish caught (and marked) in the first catch, and let N_2 be the number of fish caught in the second catch. We define the fraction, f, to be the measured fraction of the N_2 fish that are marked:

$$f = \frac{\text{number of marked fish in the second catch}}{\text{total number of fish in the second catch}}. \tag{1}$$

The total number of marked fish in the lake is known exactly—it is just N_1. The actual fraction of marked fish in the lake, which we denote by f_0, is given by $f_0 = N_1/N_0$. If the marked and released fish (the N_1 sample) mixed well with all the other fish, and the second catch

3. While this is intuitively reasonable, we show it is formally true below.

(the N_2 fish) is a random sample of the total, marked and unmarked, population (i.e., a random sample of the N_0 fish), then the measured fraction f should provide a good estimate of the actual fraction, f_0, of the fish in the lake that were marked. Hence[3] we set f equal to f_0, which then implies $f = N_1/N_0$, or

$$N_0 = \frac{N_1}{f}, \tag{2}$$

where both N_1 and f are measured quantities. Easy enough so far, but what determines the statistical error in the estimate? The only two variables at our disposal are the sizes of the two subsamples, N_1 and N_2. So we have to adjust them somehow to minimize the error. If, as is likely, we have limited time to do the censusing, so that the sum $N_1 + N_2$ is fixed, we have to find the ratio of N_1 to N_2 that will minimize the error. Thus, practically speaking, this ratio will show us the relative effort we should exert making the first catch versus making the second catch, because fishing effort is the variable that we can affect.

By the term *statistical error*, we are referring to the error that comes from random sampling. We are going to ignore another type of error—that due to systematic bias in the method. For example, the possibilities exist that a fish, once caught, is wary of being caught again (which would lead to an underestimate of the true value of f) and that some fish are prone to being caught (which would lead to an overestimate of the true f).

Because we are randomly sampling from a population of N_0 fish, of which N_1 are marked, our concern is with the uncertainty in the number of marked ones that we catch. This problem is just like pulling balls out of a jar containing a mix of red and blue balls. Because we do not care about the order in which the fish are caught, the binomial distribution described in Problem I-5 characterizes our situation. Suppose the true fraction of marked fish in the lake, N_1/N_0, is f_0; because of sampling error, this number may or may not equal the fraction f that we measure to be marked in the N_2 sample. As each fish is caught in the recapture process, the probability that any particular one of those caught fish is marked is f_0, and the probability it is unmarked is $1 - f_0$. Clearly, f_0 plays the role of p in Eq. 9 of Problem I-5.

But what is the equivalent, for our fish problem, of x and N in that equation? To draw the analogy more tightly, we divide the second catch into a marked group, $N_{2,\mathrm{m}}$ and an unmarked group $N_{2,\mathrm{u}} = N_2 - N_{2,\mathrm{m}}$. Then the quantity N in Eq. 9 is for this problem just N_2, the total size of the second catch. The variable whose probability distribution we seek, x, is the marked subset that is recaught, or $N_{2,\mathrm{m}}$. The probability of observing $N_{2,\mathrm{m}}$ marked fish is then given by

$$P(N_{2,\mathrm{m}}, N_2, f_0) = \frac{(f_0)^{N_{2,\mathrm{m}}}(1 - f_0)^{N_2 - N_{2,\mathrm{m}}} N_2!}{N_{2,\mathrm{m}}!(N_2 - N_{2,\mathrm{m}})!}. \tag{3}$$

Using Eq. 7 in I-5, we see the mean value of $N_{2,\mathrm{m}}$ is $f_0 N_2$. The observed fraction of marked fish caught in the second sampling is $f = N_{2,\mathrm{m}}/N_2$, so we have derived the result that for any fixed values of N_1 and N_2,

$$\text{the mean value of } f = f_0 = \frac{N_1}{N_0}. \tag{4}$$

So the mean value of the measured f equals the true value, f_0. Substituting this mean value of f into our formula (Eq. 2) for the estimated fish population, we get

$$\text{mean estimated fish population} = \frac{N_1}{\text{mean value of } f} = \frac{N_1}{N_1/N_0} = N_0. \tag{5}$$

Perhaps this result does not surprise you, but these kinds of statistical problems do not always work out this way—the sampling procedure might have produced bias. In a well-designed sampling procedure, our result is expected.

Now let us return to the original question: What is the error in our estimate of the population, and how can we minimize it? As we saw, this error is measured by the standard deviation, or its square, the variance, of the measured value. The estimated population is, by Eq. 2, N_1/f, where $f = N_{2,\mathrm{m}}/N_2$. So our task is to minimize the variance in $(N_1 N_2/N_{2,\mathrm{m}})$. N_1 and N_2 are numbers that we will adjust later to minimize the error, but they have no variance because they are numbers of fish caught, and we can measure them with complete accuracy just by counting carefully. From Rule 5″ in the Background to Chapter I,

$$\text{the variance in } aX \cong a^2 \text{ (the variance in } X) \tag{6}$$

and

$$\text{the variance in } \frac{1}{X} \cong \frac{\text{the variance in } X}{X^4}. \tag{7}$$

"No, it is NOT cute, and remind me never to chase jumping green flies again –they're vicious!!"

Hence,

$$\text{variance in } \frac{N_1 N_2}{N_{2,\text{m}}} \cong N_1{}^2 N_2{}^2 \frac{\text{variance in } N_{2,\text{m}}}{(N_{2,\text{m}})^4}. \tag{8}$$

From the formula for the standard deviation of a binomial distribution (Eq. 8 of I-5), the variance in $N_{2,\text{m}}$ is $N_2 f_0(1 - f_0)$. Using $f_0 = N_1/N_0$, this becomes

$$\text{variance in } N_{2,\text{m}} = N_2 N_1 \frac{N_0 - N_1}{N_0{}^2}. \tag{9}$$

Hence,

$$\text{variance in } \frac{N_1 N_2}{N_{2,\text{m}}} \cong N_1{}^3 N_2{}^3 \frac{N_0 - N_1}{N_0{}^2 N_{2,\text{m}}{}^4}. \tag{10}$$

Now, using $N_{2,\text{m}}/N_2 = f = f_0 = N_1/N_0$, we get for our best estimate of $N_{2,\text{m}}$ the quantity $N_1 N_2/N_0$, and so

$$\text{variance in } \frac{N_1 N_2}{N_{2,\text{m}}} = N_0{}^2 \frac{N_0 - N_1}{N_1 N_2}. \tag{11}$$

For simplicity, we have replaced the "almost equal" sign with an "equals" sign (but see Exercise 6). Our problem is then to find the values of N_1 and N_2 that minimize the quantity $N_0{}^2(N_0 - N_1)/(N_1 N_2)$. We must remember that there is a constraint on N_1 and N_2—their sum must not exceed a fixed constant, C, which represents how much total effort we are willing to devote to catching fish. Without that constraint, the answer is obvious—let $N_1 = N_0$, and then there is no error in the estimate. So, substituting $C - N_1$ for N_2, we now have to find the value of N_1 that minimizes the variance as given by

$$\text{variance} = N_0{}^2 \frac{N_0 - N_1}{N_1(C - N_1)}, \tag{12}$$

subject to the limiting condition that $N_1 \leq C$. To look for the minimum, we first take the derivative:

$$\frac{d(\text{variance})}{dN_1} = \frac{N_1(C - N_1)(-N_0{}^2) - N_0{}^2(N_0 - N_1)(C - 2N_1)}{[N_1(C - N_1)]^2}. \tag{13}$$

Then the extremum value for N_1 is given by setting the numerator of this derivative to zero and solving for N_1:

$$N_1(C - N_1)(-N_0{}^2) - N_0{}^2(N_0 - N_1)(C - 2N_1) = 0, \tag{14}$$

or, after rearranging and cancelling terms,

$$N_1(C - N_1) = (2N_1 - C)(N_0 - N_1). \tag{15}$$

This equation can be rewritten in the more familiar form of a quadratic equation:

$$N_1{}^2 - 2N_0N_1 + CN_0 = 0, \tag{16}$$

which has the two solutions:

$$N_1 = N_0\left[1 \pm \left(1 - \frac{C}{N_0}\right)^{1/2}\right]. \tag{17}$$

We can reject the solution with the positive sign because it implies $N_1 > N_0$, which is impossible. To simplify the other solution, let's make the generally reasonable assumption that we will not be pulling out of the lake more than a very small fraction of the fish in it, so that $C \ll N_0$. In that case, we can approximate the square root as follows (see Appendix):

$$\left(1 - \frac{C}{N_0}\right)^{1/2} \sim 1 - \frac{1}{2}\frac{C}{N_0}, \tag{18}$$

and then our solution becomes

$$N_1 \sim \frac{C}{2}. \tag{19}$$

Taking the next term in the expansion of the square root (see Appendix),

$$\left(1 - \frac{C}{N_0}\right)^{1/2} \sim 1 - \frac{1}{2}\frac{C}{N_0} - \frac{1}{8}\left(\frac{C}{N_0}\right)^2, \tag{20}$$

we get

$$N_1 \sim \frac{1}{2}C + \frac{1}{8}\frac{C^2}{N_0}. \tag{21}$$

Hence, in the limit of $C/N_0 \ll 1$, the extremum is achieved if N_1 is approximately equal to $C/2$, or perhaps a little bigger than $C/2$ if we think the total catch, C, is not very much smaller than N_0.

Suppose that the C value we select is actually comparable to N_0. After all, we do not know in advance how big N_0 is, so we may inadvertently pick a C value that is comparable to N_0. The value of C, of course, cannot exceed N_0, so let's let $C = (1 - \lambda)N_0$, where $\lambda \ll 1$. Substituting this value for C into Eq. 17 (using the minus sign), our solution becomes

$$N_1 = N_0(1 - \lambda^{1/2}). \tag{22}$$

Thus, $N_1/N_2 = N_1/(C - N_1) \sim (1 - \lambda^{1/2})/(\lambda^{1/2} - \lambda)$, which for very small λ is approximately $1/\lambda^{1/2}$. One over the square root of a number much less than 1 is much greater than 1, so in the limit of $C \sim N_0$, we should select $N_1 \gg N_2$. In other words, we should place at least as much effort into the first catch as into the second catch.

We have seen that the extremum solution for N_1 must lie between $C/2$ and C, lying near the $C/2$ value if $C \ll N_0$ and near the C value if $C \sim N_0$. In Exercise 1, you will show that this extremum actually leads to a minimum, not a maximum, error. It is always the case that it pays to put more effort into the first catch than into the second, but the difference between the two catches should be very small if $C \ll N_0$.

So this is a fine kettle of fish: To design the best census, we have to know N_0, which is what the census is trying to determine! The analysis was not a waste of time, however, because we did learn one thing—we should take $N_1 \geq N_2$. The remaining question is: By how much should N_1 exceed N_2? Clearly, if you are about to take a census of a big, productive lake with thousands of fish in it, and you only have time to catch a few dozen fish for your census, you will not go wrong putting the same amount of effort into each catch, thereby striving for $N_1 = N_2$. After doing the homework exercises below, you will gain some useful insight into the sensitivity of the error to small deviations from the optimum, and you will see that there is really only a very narrow range of values for C/N_0 that will lead you to an unnecessarily large error if you let $N_1 = N_2$.

EXERCISE 1: Show that the solution to Eq. 17 that uses the minus sign in front of the square root actually corresponds to a minimum, not a maximum, error in the estimate of N_0.

EXERCISE 2: Consider a lake with exactly 1,000 fish. For $C = 100$, 400, and 800, calculate the optimum value of N_1/N_2.

EXERCISE 3: For that same lake with 1,000 fish, and for each of the following cases, calculate from Eq. 12 the variance in the estimate of N_0:

case	N_1	N_2
1	50	50
2	60	40
3	200	200
4	225	175
5	400	400
6	550	250

What can you conclude about the sensitivity of the error to your deviation from the optimum value of N_1/N_2?

EXERCISE 4: Consider a lake with 1,000 fish in it and $N_1 = N_2 = 200$. Estimate from the binomial distribution the probability that you will get an answer of exactly 1,000 for N_0.

EXERCISE 5: Finally, for the same lake, and for the census with $N_1 = N_2 = 50$, estimate the probability that the census will yield an answer of $\geq$1,250 fish. [*Hint*: You cannot catch a fraction of a fish.]

EXERCISE 6: Try to derive the next-leading correction to the variance of $f(x)$ as given in rule 5″ of Chapter I, and use it to estimate the magnitude of the error we made (by assuming Eq. 7 to be exact) in our estimate of the variance in $(N_1N_2/N_{2,\mathrm{m}})$ as given by Eq. 11.

II-3. Blowing in the Wind

What is the maximum power that can be extracted from a wind with a steady velocity V, using a windmill whose blades sweep out an area A_0?

. . . .

First let's figure out how much power is carried by the wind itself. This amount will set an upper limit (that may or may not be achievable) on how much power can be extracted from the wind with a windmill. Then we can figure out how much of that power can be extracted by the windmill and perhaps converted to electric power or used to lift water.

Power is flow of energy. Whereas energy can be expressed in units of joules, power can be expressed in units of joules per unit time. A watt, which is a joule per second, is thus a unit of power. Consider a wind blowing at a steady velocity V, as shown in Figure II-2. At any instant, the amount of kinetic energy (KE) possessed by the moving air within the little cylinder of length L and circular opening area A_0 is one-half the mass, M, of the air in the cylinder times the square of its velocity, or $KE = (1/2)MV^2$. The mass, M, equals the density of air, ρ, times the volume of the cylinder, or $M = \rho A_0 L$. The rate of flow of energy (i.e., the power, P) is given by the kinetic energy in that cylinder divided by the time it takes to replace that mass of air with another one. That time period, t_0, is just the time it takes for a molecule of air in the wind to travel the length of the cylinder, or $t_0 = L/V$. Hence,

$$P = \frac{KE}{t_0} = \frac{1}{2}\frac{\rho A_0 L V^2}{L/V} = \frac{1}{2}\rho A_0 V^3. \quad (1)$$

If the moving air mass were not disturbed by the windmill, and its velocity on the downwind side of the windmill were also equal to V,

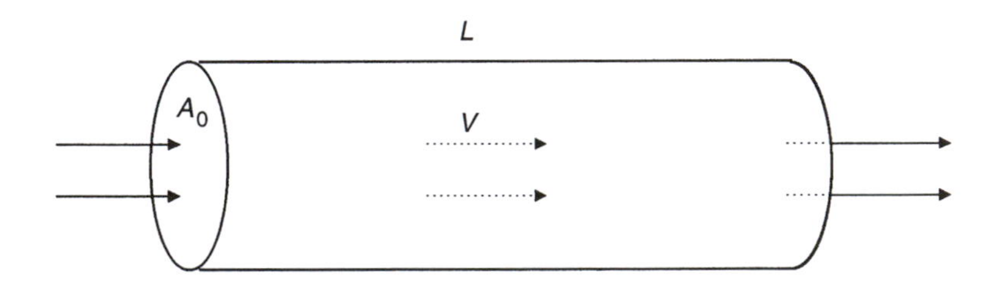

Figure II-2 Schematic of a cylinder of length L and circular opening area A_0 through which flows an undisturbed wind stream of velocity V.

then no power would be extracted. The windmill must slow down the wind if energy is to be transmitted from the wind to the windmill. If the air in the imaginary cylinder shown above is slowing down, either (a) the cylinder must widen as the velocity is reduced, or (b) the density of the air must be increased. Otherwise, the wind would be blocked by the more slowly moving air ahead. In somewhat the same way, fast-moving cars approaching a bad stretch of road can maintain their speed right up to the bad stretch only if (a) the road widens where the road is bad, or (b) the cars get sufficiently closer together on the bad stretch.

Although air is compressible, the ideal gas law tells us that the density of the air within the flow stream must remain nearly constant because the temperature and pressure of the air are not altered much by the presence of the windmill.[4] Hence, the flow path must look something like that shown in Figure II-3; the flared shape represents the outer boundaries of a patch of moving air that strikes the blades. The velocity of the wind far upstream of the windmill is V_i, where the subscript "i" refers to *initial*; far downstream, the wind velocity is V_f, where "f" refers to *final*; at the windmill, the wind velocity is denoted by V. To keep the density constant, the product of cross-sectional area times air velocity must remain constant all along the flow. Hence,

$$A_i V_i = A_0 V = A_f V_f. \qquad (2)$$

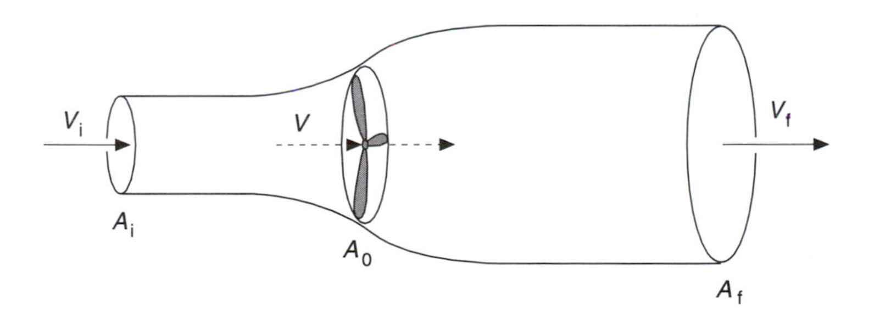

Figure II-3 Schematic representation of a wind stream in the presence of a windmill blade; the flared shape represents the outer boundaries of a patch of moving air that strikes the blades. V_i, initial wind velocity upstream of the windmill; V, wind velocity at the windmill; V_f, final wind velocity downstream of the windmill.

4. This law states $PV = nRT$, where P is the pressure, V is the volume of a patch of air, T is the air temperature, n is the number of moles of air in the volume, and R is the famous ideal gas constant (see Appendix II in *COW-1*). Using the fact that density is proportional to n/V, we see that constant T and P implies constant density.

Armed with these preliminaries, we can calculate the efficiency, e_w, of extraction:

$$e_w = \text{fraction of wind power extracted by windmill}$$
$$= \frac{\text{power extracted from the wind by windmill blade}}{\text{power in a cross section of wind of area equal to that swept out by windmill blade}}. \quad (3)$$

The denominator is just the expression $(1/2)\rho A_0 V_i^3$ derived above. Note that we take the area of the incoming wind to be that swept out by the windmill blade, but the velocity of the wind is not that at the blade itself but rather the undisturbed, or incoming, velocity.

To calculate the numerator, we need to briefly review Newton's laws of motion and their consequences. The momentum of a moving object is its mass times its velocity. The force exerted by such an object when it strikes another object is equal to the rate of loss of its momentum due to the impact. For our moving "object," we take the quantity of air that passes by the windmill blade in a unit of time (say one second). During that second, the quantity of air passing the windmill is equal to $\rho A_0 V$. We will call that a "unit" of air. The magnitude of the change in the velocity of that unit of air is $V_i - V_f$. Thus, the force exerted on the windmill blade, or the change in momentum of that unit of air, is given by the mass of the unit of air times the change in its velocity:

$$\text{force on blade exerted by unit mass of air} = \rho A_0 V (V_i - V_f). \quad (4)$$

Another fundamental concept from the physics of motion concerns energy in the form of work. Work is a manifestation of the energy incorporated into the motion of objects; work is done when a force moves an object. The amount of work done by an object exerting a force F when the object moves a distance L is equal to FL. During one time unit, the unit air mass moves a distance V past the blade, so the work done by the wind against the blade is

$$\text{work done on blade by unit mass of air} = \rho A_0 V^2 (V_i - V_f). \quad (5)$$

This expression was derived from consideration of momentum and force, but now we can write yet another exact expression for "work done on the blade per unit mass of air hitting blade," based on consideration of the change in kinetic energy in the wind. The kinetic energy of a mass M of air moving at a velocity V is $MV^2/2$, and the change in the kinetic energy of the wind is equivalent to that embodied as the work done by the wind on the windmill blade. For a unit of air mass, $\rho A_0 V$, moving past the blade,

$$\text{the loss in kinetic energy in unit mass of air} = \frac{\rho A_0 V(V_i^2 - V_f^2)}{2}. \quad (6)$$

Equating these two expressions (Eqs. 5 and 6) for the same thing, we get

$$V = \frac{V_i + V_f}{2}. \quad (7)$$

The power extracted from the wind by the turbine can be obtained from either the expression for loss of kinetic energy (Eq. 6) or the expresson for work done on the windmill blade (Eq. 5). The power extracted is the rate at which energy is extracted. The amount of energy extracted in a unit of time is equal to the energy extracted by the unit of air mass divided by the time it takes for the mass of air to pass the blade. Because the unit air mass takes, by definition, one unit of time to pass the blade, the power extracted is given by Eq. 6. Hence, the efficiency can be reexpressed, using Eq. 7, as

$$e_w = \frac{\dfrac{\rho A_0 V(V_i^2 - V_f^2)}{2}}{\dfrac{\rho A_0 V_i^3}{2}} = \frac{V(V_i^2 - V_f^2)}{V_i^3} = \frac{\dfrac{(V_i + V_f)(V_i^2 - V_f^2)}{2}}{V_i^3}. \quad (8)$$

Next, we introduce a parameter, b, that we can vary to find the maximum efficiency. To this end, we let $V_f = bV_i$. The efficiency then becomes

$$e_w = \frac{\dfrac{V_i^3(1+b)(1-b^2)}{2}}{V_i^3} = \frac{(1+b)(1-b^2)}{2} = \frac{1+b-b^2-b^3}{2}. \quad (9)$$

Our task now is to find the value of b that maximizes this expression, which we do by first setting $de_w/db = 0$: $de_w/db = (1 - 2b - 3b2)/2 = 0$. This leads to the quadratic equation: $3b^2 + 2b - 1 = 0$, or $b = [-2 \pm (4+12)^{1/2}]/6 = 1/3$, or -1. The root $b = -1$, when substituted back in to Eq. 9, yields $e_w = 0$, which is clearly a minimum, not a maximum. Substituting $b = 1/3$, we get $e_w = (4/3)(8/9)/2 = 16/27$.

Thus, the maximum efficiency is 16/27. This is called the Betz limit —it asserts that a windmill can extract from the wind at most a fraction (16/27) of the wind power incident on it in a cylindrical flow of air with diameter equal to that of the windmill blade. Real windmills will not function anywhere near this efficiently if the blades are poorly designed, if they have considerable internal friction, or if many windmills are packed closely together so that they interfere with each others' wind streams, either upwind or downwind.

EXERCISE 1: What is the power in the flowing water at the base of a 10-m high waterfall, where the water has a cross sectional area A and water velocity V? What is the maximum fraction of this power that could be extracted by a water turbine? (Careful, the factor of 16/27 is no longer valid—do you see why?)

EXERCISE 2: Because the power in the wind is proportional to the cube of the wind speed, estimating the wind resource requires care. Consider a varying wind, with an average speed of 10 km/h and a uniform probability of speeds within the range of 0 km/h to 20 km/h. Compare the cube of the average wind speed to the average value of the cube of the wind speed. Which result provides the correct value to insert in our formula for the power flux in the wind?

EXERCISE 3: For the denominator in Eq. 3 (i.e., the available power in the wind), we took the power in a cross section of wind of area equal to that swept out by the windmill blades. In reality, taking that area to be A_f might be more reasonable because, if windmills are located so close to one another that the A_f circles of adjacent windmills overlap, then the windmills will be interfering with their neighbors' air streams. How does this substitution modify the Betz limit?

EXERCISE 4: Assuming windmills can generate electricity at the maximum efficiency derived above (i.e., the Betz limit) and using a reasonable estimate of the wind resource, estimate how much land area dedicated to windmills would be needed to produce enough electricity to meet the current US electricity demand. To what fraction of the US land area does your answer correspond? What are some of the factors you would want to consider before promoting wind as the "answer" to the problem of meeting US electricity needs?

II-4. Haste Makes Waste

The maximum efficiency of a Carnot engine, operating between a heat source at temperature T_H and a heat sink at temperature T_C, is given by the famous formula: $e_{max} = (T_H - T_C)/T_H$. An optimally efficient Carnot engine must operate reversibly. In practice, however, getting heat to a Carnot engine and removing the waste heat it produces cannot be achieved reversibly unless these energy flows are infinitely slow.[5] Hence, the ideal Carnot process is, practically speaking, a fiction. Here you will look at the more realistic case of an optimally efficient irreversible engine. As before, we have a heat source at fixed temperature T_H and a sink at fixed temperature T_C (see Figure II-4). Heat from this source to the engine, and from the engine to this sink, is transferred by conduction (an irreversible process) at rates F_H and F_C, respectively. The rate F_H is proportional to the temperature difference $T_H - T_2$, and the rate F_C is proportional to the temperature difference $T_1 - T_C$. A traditional reversible Carnot engine operates between the resulting hot temperature T_2 and the cold temperature T_1, so its maximum efficiency is $e_{max} = (T_2 - T_1)/T_2$, with temperature measured on the Kelvin scale. The temperatures T_2 and T_1 are subject to control by the designer and in particular can be adjusted in such a way as to maximize the power output of the process. Derive a formula for the maximum power output and the corresponding efficiency of this irreversible process in terms of only T_H and T_C.[6]

. . . .

---

5. Readers lacking a good intuitive understanding of thermodynamics, and in particular the connection between reversibility, infinitely slow processes, and maximum efficiency, are urged to look at Van Ness' superb book: Van Ness, H. C. 1969. *Understanding thermodynamics*. New York: McGraw-Hill.

6. The basis of this problem is found in Curzon, F., Ahlborn, B. 1975. Efficiency of a Carnot engine at maximum power output. *Am. J. Physics* 43:22–24.

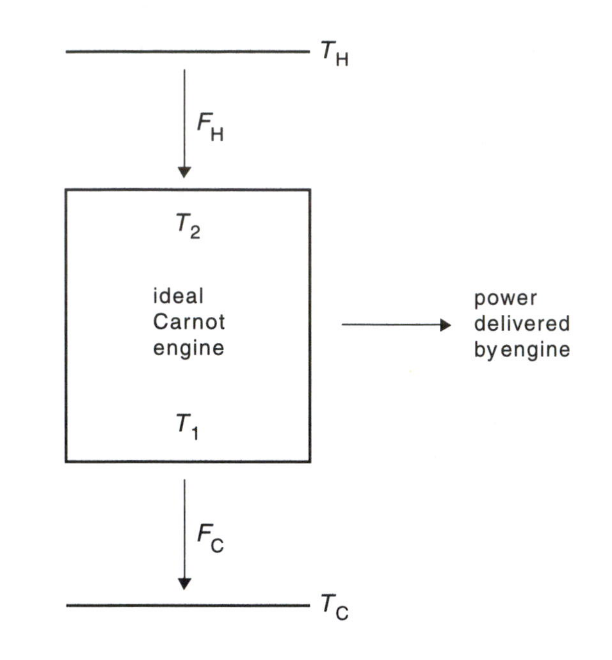

Figure II-4 An irreversible Carnot process. A reversible, ideal Carnot engine operates between the two temperatures, T_2 and T_1, but the transfer of heat from T_H to T_2, at rate F_H, and from T_1 to T_C at rate F_C, is irreversible. T_H and T_C should be thought of as fixed by the constraints of fuel type and the environment into which waste heat is dumped. For example, T_H might be the temperature of the fire heating the working fluid, and T_C might be the temperature of the river water passing through the cooling condensers. The efficiency of the ideal Carnot engine is given by $(T_2 - T_1)/T_2$, and T_1 and T_2 can be adjusted by the system engineer in such as way as to maximize power output.

By the principle of conservation of energy, the power delivered by this engine is just the difference between the rate of energy delivered to the engine, F_H, and that delivered to the heat sink in the waste heat stream, F_C. Thus,

$$P_{out} = F_H - F_C. \tag{1}$$

Our task is to maximize this power output, by adjusting T_1 and T_2 and then to evaluate the efficiency of the ideal Carnot engine,

$$e_{Carnot} = \frac{T_2 - T_1}{T_2} \tag{2}$$

for this optimal case. The relevance of this ideal Carnot engine efficiency to the operating characteristics of the actual heat engine may not be obvious to you now, considering that the ideal Carnot engine is only a part of the full irreversible system, but its relevance will become clear later (Exercise 3).

Because the conductive flows of energy are proportional to the temperature differences, we can write

$$F_{\mathrm{H}} = k(T_{\mathrm{H}} - T_2) \tag{3}$$

and

$$F_{\mathrm{C}} = k(T_1 - T_{\mathrm{C}}). \tag{4}$$

Note that we have taken the conductivity constant, k, to be the same for both conduction processes, but in Exercise 5 you will see that we could just as well have let them differ. Substituting Eqs. 3 and 4 into Eq. 1, we can express the power output in terms of the temperatures:

$$P_{\mathrm{out}} = k(T_{\mathrm{H}} + T_{\mathrm{C}} - T_2 - T_1). \tag{5}$$

The next step is to eliminate either T_2 or T_1 from this expression so that we can maximize it with respect to the remaining variable. Recall that T_{H} and T_{C} are not adjustable.

From traditional reversible thermodynamics, when the Carnot engine in the center of this process is maximally efficient,

$$\frac{F_{\mathrm{H}}}{F_{\mathrm{C}}} = \frac{T_2}{T_1}. \tag{6}$$

In Eq. 6, and throughout the remainder of this problem, we need to use the Kelvin scale to express temperatures. Combining Eq. 6 with Eqs. 3 and 4, we get

$$T_1 = \frac{T_2 T_{\mathrm{C}}}{2T_2 - T_{\mathrm{H}}}. \tag{7}$$

Substituting Eq. 7 into Eq. 5,

$$P_{\mathrm{out}} = k\left[T_{\mathrm{H}} + T_{\mathrm{C}} - T_2 - \frac{T_2 T_{\mathrm{C}}}{2T_2 - T_{\mathrm{H}}}\right], \tag{8}$$

and thus we have achieved our first goal of obtaining P_{out} as a function of a single adjustable variable, T_2.

Now we can maximize this expression, by setting $dP_{\mathrm{out}}/dT_2 = 0$:

$$\frac{dP_{\mathrm{out}}}{dT_2} = k\left[-1 - \frac{T_{\mathrm{C}}}{2T_2 - T_{\mathrm{H}}} + \frac{2T_2 T_{\mathrm{C}}}{(2T_2 - T_{\mathrm{H}})^2}\right] \tag{9}$$

and so the maximization condition becomes

$$1 + \frac{T_C}{2T_2 - T_H} = \frac{2T_2T_C}{(2T_2 - T_H)^2}. \tag{10}$$

Using a little algebra, this expression can be rewritten as

$$T_2 = \tfrac{1}{2}[T_H \pm (T_C T_H)^{1/2}]. \tag{11}$$

Substituting into Eq. 7, we get

$$T_1 = \tfrac{1}{2}[T_C \pm (T_C T_H)^{1/2}]. \tag{12}$$

Which sign in front of the square root corresponds to the maximum power output? By taking second derivatives, or plotting P_{out} versus T_2, you can show (Exercise 2) that the positive sign is the one we want; moreover, the choice of negative sign would imply a negative T_1 temperature.

Now we can substitute Eqs. 11 and 12 into Eq. 2 to get the efficiency of the Carnot process at this maximum rate of power production:

$$e_{max} = \frac{T_2 - T_1}{T_2} = \frac{\sqrt{T_H} - \sqrt{T_C}}{\sqrt{T_H}}. \tag{13}$$

This equation is like the expression for the traditional Carnot efficiency, except that the square roots of T_H and T_C appear instead of the temperatures themselves.

EXERCISE 1: Is the efficiency in Eq. 13 greater or less than the ideal Carnot efficiency for a reversible heat engine? Does your answer make sense?

EXERCISE 2: Show that the positive signs in Eqs. 11 and 12 actually correspond to the maximum value of P_{out}.

EXERCISE 3: This exercise gives you insight into the actual relevance of the ideal Carnot efficiency $(T_2 - T_1)/T_2$ to our irreversible system. First, substitute the optimum value for T_2 (Eq. 11) into Eq. 8, to derive a formula for the maximum power output in terms of T_H and T_C. Then express this power output as a product of e_{max} (as given by Eq. 13) and some power input, P_{in}, and show that $P_{in} = F_H$. Explain why this result justifies our identifying e_{max} with the actual conventionally measured efficiency of the power plant.

EXERCISE 4: The results from Exercise 3 above justify identifying the efficiency in Eq. 13 with the optimum efficiency obtained in real power plants. Modern coal-fired, electric-generating power plants operate at a maximum efficiency of approximately 40% and between a heat source of temperature $T_H \sim 840$ K and a heat sink of temperature

$T_C \sim 300$ K. Geothermal steam facilities generally produce electricity at a maximum efficiency of 17%, with $T_H \sim 520$ K and $T_C \sim 350$ K. For coal-fired and for geothermal power generation, what are the optimum reversible-process and irreversible-process efficiencies under these temperature conditions? Compare your answers to the actual efficiencies.

EXERCISE 5: Show that Eq. 13 still holds even if the conductivity coefficients are different at the heat source and heat sink. In other words, replace Eqs. 3 and 4 with $F_H = k_H(T_H - T_2)$ and $F_C = k_C(T_1 - T_C)$, where $k_H \neq k_C$, and rederive Eq. 13.

II-5. Biting the Hand that Feeds Us (I)

Economic activities nearly always result in some degree of ecological degradation. At the same time, healthy ecosystems carry out a host of processes, providing what are sometimes called ecosystem services,[7] that contribute to our well-being and help sustain our economies. Figure II-5 illustrates how this two-way relationship between ecosystems and economic activity is imbedded within the larger context of human well-being and how it implies a tradeoff between the gain in human well-being resulting from economic production and the loss caused by ecosystem degradation. Make a simple model of this situation and see what it tells you about the optimum level of economic activity.

. . . .

To see the implications of this two-way interaction between economies and ecosystems, let's start with what may be the simplest possible model[8] that captures the essential relationships among economic activity, ecological integrity, economic output, and human well-being in a sustainable world. I use the term *sustainable* here to mean persistent over many human lifetimes, but not necessarily desirable; a lifeless earth may well be persistent. The time frame for which this simple model makes the most sense is on the order of decades—a frame that is intermediate between the interannual scale of weather and economic fluctuation and the much longer time spans over which ecological succession and major social and technological change and innovation need to be considered. The spatial scale of analysis could be as small as a rural village or county, or as large as a nation or the globe.

In Figure II-5, economic activity (X) and economic output (Y) are distinguished. X is the level of those economic activities that reduces environmental goods and services; it refers to such measures of activity as the rate of fossil fuel consumption, the number of cattle being

---

7. These services include such benefits as dampening of climatic and hydrologic extremes, formation and maintenance of fertile soil, natural pest control, removal of pollutants from water, air, and soil, and maintenance of a diverse gene pool from which we derive seeds for agriculture and cures for many diseases.

8. In a subsequent problem (V-3), we will extend the analysis by adding more realism to the simple model developed here. Here, the model's simplicity means we can really only look at sustainable or steady-state futures, but in Problem V-3, we will examine time-varying future trajectories.

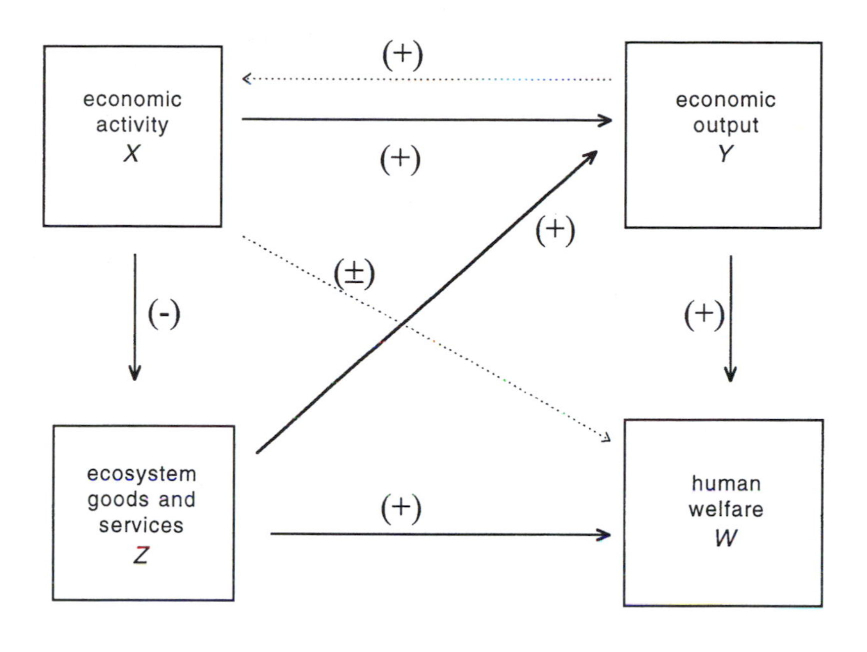

Figure II-5 Highly simplified representation of the major interdependencies in the economic–ecologic system. Arrows indicate the primary direction of influence associated with each interrelationship. A plus (minus) sign indicates the box at the tail of the arrow enhances (depresses) the box at the head. Only the solid lines are considered in this problem.

grazed, and the annual tonnage of minerals being mined. It is thus a composite measure of total economic activity that to one degree or another occurs at the expense of environmental quality. I emphasize that X refers to total, not per-capita, activity. Y refers to the total output of an economy—for a nation, it would be the gross economic product. Y can be measured directly in monetary units. Z, the quality of the environment, is a composite measure that includes, for example, air and water quality, biodiversity, soil fertility, integrity of hydrologic processes such as natural flood control, and the aesthetic characteristics of the environment that contribute to the overall quality of the "wilderness experience" enjoyed by hikers. Finally, W is a measure of the component of total human welfare that is provided by Y and Z.

Figure II-5 illustrates three dominant linkages:

1. Welfare is positively influenced by economic output and environmental quality.
2. Economic output is positively influenced by economic activity and environmental quality.

3. Environmental quality is negatively influenced by economic activity.

A fourth linkage—a positive influence of economic output on economic activity (e.g., investment of profits from recreation into housing and new tourist facilities)—is shown as a dashed line and will not be treated here. The reason is that we treat economic activity as an independent, or exogenous, "planned" variable. A fifth linkage, also shown as a dotted line and neglected here, corresponds to the influences on welfare that result directly (that is, not via economic output or ecological degradation) from increasing economic activity. These influences can be negative (e.g., job stresses, neglect of families) or positive (e.g., pride in work skill). Although the figure conveys the impression that we are lumping economic activities that have large impacts on the environment (e.g., coal burning) with those that achieve the same goal but are relatively innocuous (e.g., use of solar energy), such distinctions can be readily incorporated into the framework provided.

To explore the consequences of the linkages highlighted in the figure, we construct a mathematical representation of the pattern. Given the relative lack of detailed knowledge about mechanisms and relationships, many such representations can be written in the form of sets of equations. In the interests of keeping the equations simple yet plausible, we suggest the following mathematical model as a starting point for further analysis:

$$W = aYZ \tag{1}$$

$$Y = bXZ \tag{2}$$

$$Z = Z_0 - cX. \tag{3}$$

"Finish your lunch already! I don't want to sit around here all day!"

The first two equations are algebraic descriptions of instantaneous relationships among the model components. Eq. 1 states that total human welfare depends on both economic output and environmental quality. I have chosen a product form, rather than an additive form, $\alpha Y + \beta Z$, because we assume that in the limit of complete deterioration of the economy *or* the environment, welfare would be effectively as reduced as if both so deteriorated. This choice reflects an implicit judgment about the zero points of the metrics with which we measure welfare, output, and quality. The product form contrasts with many models of human welfare based on additive contributions from various inputs such as agricultural and industrial production and recreation.

Eq. 2 captures the notion that economic output results from a product of environmental infrastructure and economic activity.[9] As with the expression for welfare, the underlying assumption is that if either activity or quality diminishes to zero, then so does output. An additive form, in contrast, would be based on the very different assumption that activity alone, or environmental quality alone, could result in economic output. The constant b, multiplying XZ in this expression, is a surrogate for a complex array of factors such as multiplier effects, external subsidies, and other economic and social ingredients that influence economic output. An underlying assumption in the application of this equation is that the factors buried in the constant b remain fixed as we explore the effects of altering the explicit variables in the model. Another assumption in both Eqs. 1 and 2 is that changes in environmental quality result in instantaneous changes in welfare and output. There will, of course, be time lags, but we are assuming they can be neglected. Eq. 3 expresses environmental quality as a function of economic activity. In the absence of economic activity, past and present, $X = 0$, and environmental quality would be at a level Z_0. Eq. 3 also ignores the effect of a time lag in the response of environmental quality to increased economic activity, but see problem V-3.

Eqs. 1 through 3 can be used to determine the value of economic activity, X_{max}, that maximizes welfare, W, or economic output, Y. Substituting Eq. 3 into Eq. 2, and the resulting value for Y from Eq. 2 into Eq. 1, we get $W = abX(Z_0 - cX)^2$. Maximizing this expression with respect to X yields $X_{max} = Z_0/3c$ and $W_{max} = (4ab/27c)Z_0{}^3$. Although this result, by itself, is not very enlightening, Exercises 1 through 3 below will provide some insight into its relevance.

9. The product form of Eq. 2 can be altered to resemble, superficially, the form of the Cobb–Douglas production function of neoclassical economics, which relates economic output to a product of powers of the contributions of labor, resources, and capital. Thus, we could write $Y = bX^{\alpha}Z^{\beta}$, where α and β are empirically determined parameters. Our preliminary analysis is confined to the case in which $\alpha = \beta = 1$, but see Exercise 2. I emphasize that in this model, in contrast to the Cobb–Douglas formulation, the inputs of capital, labor, and resources are implicitly buried in the independent control variable X.

EXERCISE 1: Suppose that instead of maximizing welfare, W, we set the goal to be finding the value of economic activity, X, that maximizes economic output, Y. Show that the maximization of economic output leads to exactly 3/2 the level of economic activity that would maximize welfare. Discuss the implications of the fact that this ratio of output-maximizing activity to welfare-maximizing activity exceeds 1.

EXERCISE 2: How robust is the result (from Exercise 1) that the output-maximizing level of economic activity exceeds the welfare-maximizing level of activity? To explore this question, consider a wider class of models, in which $W = aY^{\rho}Z^{\lambda}$ and $Y = bX^{\alpha}Z^{\beta}$, and work out the ratio of the level of economic activity that maximizes economic output to the level that maximizes welfare.

EXERCISE 3: Under the same assumption as in Exercise 1, namely that economic output, not human welfare, is maximized, let's estimate the "cost of greed." In particular, show that the level of welfare that results from maximizing output is reduced by a factor of 27/32 relative to that which results from maximizing welfare itself. How robust is the result that this factor is less than 1?

EXERCISE 4: Show that the level of economic activity, X, that maximizes the ratio of W to Y (welfare per unit of economic output) is $X = 0$. Is that still true if the parameters ρ, λ, α, and β in Exercise $2 \neq 1$?

EXERCISE 5: Explore the consequences of adding to the model either or both of the linkages shown by dotted lines in Figure II-5.

Chapter III
Scaling and Dimensional Consistency

Gonzalo: "Now would I give a thousand furlongs of sea for one acre of barren ground."

Introduction

There are fascinating patterns in nature that enable us to extend our knowledge across space, time, mass, and energy. For example, we can predict accurately the metabolic rates of large mammals based on knowledge of the metabolic rates of small mammals. We can draw reasonable conclusions about how many plant species to expect if we could census a large patch in a forest, based on knowledge of how many species there are in small patches. We can estimate the frequency of large earthquakes or forest fires based on knowledge of the frequency of smaller ones. We can estimate the air resistance slowing a large meteorite if we know its value for a small one.

These are all examples of applications of scaling laws. Such laws give us the ability to extend our knowledge of some structure or process from one domain, where we know a good deal, to some other domain, where we have less information—they enable us to scale up or scale down our knowledge. The domain over which we scale may be space, as in the example of species richness in big and small census plots, but it may also be body weight, as in the metabolic scaling law, or time, as in the estimation of earthquake frequency.

The species-richness-versus-area relationship, sometimes called a species-area relationship, tells us that the number of species in a plot of area A scales like A^z, where z is some constant; in practice, the value of z is often around one-fourth (see Rosenzweig, M. 1995, in Scaling section of Further Reading in Appendix). The metabolic scaling relationship tells us that the basal (or resting) metabolic rates of mammals scale approximately as body weight to the three-quarters power;[10]

10. This is known as Kleiber's law. The empirical basis for the law is discussed in Schmidt-Nielsen, K. 1972. *How animals work.* London: Cambridge Univ. Press. An elegant explanation of Kleiber's law based on the assumption that the circulatory system in mammals possesses a fractal structure has been given by West, G., Brown, J., Enquist, B. 1997. A general model for the origin of allometric scaling laws in biology. *Science* 276:122–25.

A problem of scale.

thus, knowing that adult humans have a basal metabolic rate of about 2×10^6 calories per day (2,000 Calories/day in the more familiar weight-watcher's units, where 1 Calorie = 1,000 calories), I can estimate the basal metabolic rates for shrews and elephants. We will explore such relationships in this chapter's problems.

To be more specific about space and time scales, it helps to distinguish two facets of space or time: grain and interval. *Grain* refers to the resolution with which we view a phenomenon. Suppose we are interested in how the list of species of soil-dwelling insects differs from site to site within a forest. To measure this difference, we might conduct an insect survey in small patches of soil (perhaps 10 cm × 10 cm patches) at various locations scattered throughout the forest. The ratio of the number of species found in common to two such patches to the average number of species in each of the two patches is a useful measure of the uniformity of the species distribution; one minus that fraction is called *turnover*. In general, that fraction will be a function of the distance between the two patches and the areas of the patches. If we choose small patches to sample, then we are examining species turnover at a fine spatial grain. If we choose patches that are far apart, then we are studying turnover over a large *interval*.

We might observe a simple, systematic pattern in the dependence of turnover on the two variables: patch area (grain) and interpatch separation (interval). But if the patch area is too small, then we might see a "noisier" pattern because the insects living in very small soil samples might vary randomly because of the vagaries of where tree branches recently fell. Similarly, the simple pattern may hold only over a limited range of interpatch distances. For example, if the intervals studied are too great, then climatic differences between distant patches might result in a greater turnover than would be predicted from our knowledge of the turnover between patches that are closer together.

The main goal of this chapter is to familiarize you with some of the famous examples of scaling patterns in nature and have you discover some of their interesting applications. I hope you will also gain some understanding of the domains in which they hold and the boundaries beyond which they break down, and of possible mechanisms that can generate scaling patterns.

Background

Dimensional consistency. The most basic tool used in the analysis of scaling is dimensional consistency. Measurements made with clocks, rulers, and balances tell us about time, distance, and mass, respectively. We refer to time, distance, and mass as dimensions. We express measurements of these fundamental dimensions in units such as seconds, meters, and kilograms. But those units are only a convention. So physical laws, and the form of expressions describing patterns, should be independent of the units with which we express the fundamental dimensions of time, distance, and mass. Suppose we define a di-meter to be two meters and re-express the results of measurements in units of seconds, di-meters, and kilograms. Then we should still discover that planetary orbits sweep out equal areas in equal times, the period of a pendulum should still be independent of its mass, and, if a species-area relationship such as "species richness grows like the one-fourth-power of area" is verified with area measured in square meters, it should still be valid with area measured in square di-meters.

Dimensional consistency simply means that the equations describing laws and patterns have the same combinations of dimensions on the left and right sides of the equal sign. In his state of frantic fear brought on by an ocean tempest, Gonzalo may be forgiven for forgetting this, but you will not be. Consider the equation for the period of a pendulum swinging in a vacuum at the Earth's surface:

$$\text{period of pendulum} = 2\pi\left(\frac{\text{length of pendulum}}{\text{acceleration of gravity}}\right)^{1/2}.$$

The period of the pendulum has the dimension of time, T. Its length has the dimension of distance, L. The acceleration of gravity at the Earth's surface, often denoted by the symbol g, has the units of L/T^2. Let's check our expression above for dimensional consistency:

The dimension of the left side is T.
The dimension of the right side is $[(L)/(L/T^2)]^{1/2} = T$
So consistency is achieved.

This exercise might appear rather trivial, but in actuality dimensional consistency can be a useful tool for generating the form of physical laws. Again using the pendulum example, suppose you could not recall the formula for the period. If you made a list of all the things the period might possibly depend upon, your list would probably consist of the acceleration of gravity (g) and the length and mass of the pendulum. How can we form a quantity with dimensions of time from g, length, and mass? Consider an expression for the period as an arbitrary product of powers of these entities: $[g^a][\text{length}^b][\text{mass}^c]$. Letting M be the dimension of mass, the dimensions of this expression are $[(L/T^2)^a][L^b][M^c] = [L^{a+b}][T^{-2a}][M^c]$. For dimensional consistency, this expression must have units of T^1. Hence, $a + b = 0$, $a = -1/2$, and $c = 0$. From this, we deduce that $b = 1/2$. Substituting these values back into the original expression, we can conclude that the period is independent of mass ($c = 0$) and must be proportional to $(\text{length}/g)^{1/2}$. The only thing we could not deduce from dimensional consistency was the factor 2π in front. We shall explore less trivial applications of dimensional consistency in Problems III-1, III-3, and III-5.

You might wonder why we picked a product of powers of the entities acceleration of gravity, pendulum length, and pendulum mass—why not more complicated expressions like log(length) or e^{mass}? The reason is that you cannot consistently express a physical quantity with such functions. The quantity e^x can be written (see Appendix) as an infinite power series $1 + x^1/1! + x^2/2! + x^3/3! + \cdots$. If x has dimensions of M, then each term in this series has different dimensions, namely, ever higher powers of M. The only entity that can appear as the argument of a function such as the exponential or the logarithm is a dimensionless entity—a pure number. The only dimensional expressions involving M, L, and T that are legitimate in physical laws are sums of products, and each term in the sum has to have the same dimensions.[11]

Power laws and self-similarity. Our focus here is on a class of patterns observed by scientists studying a variety of phenomena—patterns that are well described by power-law relationships. A power-law relationship is of the form $Y = cX^z$; z is referred to as the exponent. We referred above to two examples of such power-law patterns—the metabolic scaling relationship, (metabolic rate) $\sim$ (body mass)$^{3/4}$, and the species-area relationship, (number of species) $\sim$ (area)$^{1/4}$. We emphasize that the exponents, $3/4$ and $1/4$, are not integers and that

---

11. Further discussion of dimensional consistency and a more rigorous discussion of the reason the dimension function must always be a sum of products of powers is given in Barenblatt, G. I. 1996. *Scaling, self-similarity, and intermediate asymptotics*. London: Cambridge Univ. Press. This text is an excellent source of more advanced information on many of the topics treated in this chapter.

12. For example, the oft-cited example of the self-similarity of the shape of coastlines at best makes sense at spatial scales bounded between the contintental and the molecular.

the power-law relationships we are interested in here are not in any way related to our analysis of dimensional consistency, in which products of *integral powers* of the fundamental dimensions M, L, and T are constructed. Power-law relationships with noninteger exponents are observed to describe a variety of patterns found in nature, from storm, wildfire, or earthquake frequency (for which the relationship is that the likelihood of a rare event has an inverse power dependence on the magnitude of the event), to the dependence of instream water flow on stream-channel characteristics.

The mathematics of self-similarity provides another way to describe patterns. A pattern is said to be *self-similar* if it looks the same at different spatial scales. Strictly speaking, no such patterns exist in nature across *all* scales.[12] But evidence for limited self-similarity does exist in fields ranging from geology to hydrology to ecology. Interestingly, there is a deep connection between self-similarity and power-law relationships, which I now illustrate for the specific case of the species-area law.

Consider a rectangular landscape of area A_0, with length and width in the ratio of $\sqrt{2}$, and populated with a variety of plant species. If the landscape is repeatedly bisected perpendicular to the long dimension, then at each bisection the two smaller rectangles that result also have a length-to-width ratio of $\sqrt{2}$. At the ith bisection, denote the area of the resulting rectangles by A_i, where $A_i = A_0/2^i$. Suppose the individuals within the various plant species are distributed across this landscape according to the following rule: *of the species found in each of the* A_i *rectangles, on average a fraction,* a, *are also found in at least the left-half* A_{i+1} *rectangle formed from bisecting the* A_i *rectangle.* Similarly, a fraction a are found in at least the right-half A_{i+1} rectangle. The phrase "at least" means that $\boldsymbol{a}$ is the probability that an arbitrary species is in the left half, irrespective of whether or not it is also in the right half, and vice versa.

Because the parameter $\boldsymbol{a}$ is independent of the index i and thus independent of area, the distribution of all the species is *self-similar*; that is, the rules governing the distribution of species are the same at every spatial scale—they are scale-independent. The rule governing the distribution of species applies to an "average" A_i rectangle but not necessarily to any particular one, so the distribution is *statistically*, not deterministically, self-similar. By the comment above about self-similarity holding only over a limited range of scales, we might expect that if the bisection process is carried too far (for example, down to such small scales that a single plant can no longer fit in an A_i rectangle for large values of i), then self-similarity should break down.

We show now that the species within this self-similar landscape obey the power-law form of the species-area relationship (number of species $\sim$ area^z); moreover, we derive an explicit connection between the parameter z in the species-area relationship and the parameter, $\boldsymbol{a}$, describing self-similarity.[13] Within the entire landscape, assume that there are S_0 species. By the probability rule, in each half of the landscape, there are, on average, $\boldsymbol{a}S_0$ species. In each quarter of the landscape (half of a half) the rule tells us that there are, on average, $\boldsymbol{a}^2S_0$ species, and so on. Thus, letting S_i be the average number of species in an area A_i,

$$S_i = S(A_i) = S_0\boldsymbol{a}^i. \tag{1}$$

Now let $z = -\log_2(\boldsymbol{a})$, or $\boldsymbol{a} = 2^{-z}$. Then Eq. 1 becomes

$$S(A_i) = S_0 2^{-iz}. \tag{2}$$

But $2^{-i} = A_i/A_0$, so Eq. 2 can be rewritten as

$$S(A_i) = S_0\left(\frac{A_i}{A_0}\right)^z \tag{3}$$

or

$$S(A) = \frac{S_0}{{A_0}^z}A^z. \tag{4}$$

Hence, setting $(S_0/{A_0}^z)$ equal to a constant c, we get the famous power-law form of the species-area relationship:

$$S(A) = cA^z, \tag{5}$$

where the parameter z is related to the parameter $\boldsymbol{a}$ by the equation

$$z = -\log_2(\boldsymbol{a}). \tag{6}$$

Other tricks and tools can be used to study scaling, but none is as pervasive as dimensional analysis and power-law analysis. I will introduce these other methods as we need them in the problems that follow.

---

13. The derivations presented here and in Problem III-5, and further exploration of the implications of self similarity in ecology, are discussed in a series of papers:

Harte, J., Kinzig, A. 1997. *On the implications of species-area relationships for endemism, spatial turnover, and food web patterns. Oikos* 80:417–27.

Harte, J., Kinzig, A., Green, J. L. 1999. *Self-similarity in the distribution and abundance of species. Science* 284:334–36.

Harte, J., McCarthy, S., Taylor, K., Kinzig, A., Fischer, M. 1999. *Estimating species-area relationships from plot to landscape scale using species spatial-turnover data. Oikos* 86:45–54.

Banavar, J., Green, J. L., Harte, J., Maritan, A. 1999. *Finite size scaling in ecology. Phys. Rev. Lett.* 83:4212–14.

EXERCISE 1: Of the S_i species found in an A_i rectangle, what is the fraction found *only* in the left-half A_{i+1} rectangle formed from bisection? Or, equivalently, what is the fraction found *only* in the right half? Express your answer in terms of the parameter a. What fraction of the species found in an A_i rectangle is found in *both* the left and right halves?

EXERCISE 2: A species is considered endemic to an area A if the species is found only in that area. Using your answer to Exercise 1, derive an "endemics-area relationship" in the power-law form:

$$E(A) = c'A^{z'}.$$

What is the relation between z' and z? Explain in words why z in the species-area relation (Eq. 5) must be less than or equal to 1, whereas z' in the endemics-area relationship must be greater than or equal to 1.

EXERCISE 3: In Equation 4, we showed that self-similarity implies the power-law form of the species-area relationship. Does the logic work in reverse? That is, is every species distribution that obeys the power-law form of the species-area relationship self-similar?

"OK, everybody, listen up, 'cause for the next 2 weeks those biologists expect to see us in certain places. . . "

Problems

III-1. How Tall Can a Mountain Be?

Consider a cylindrical mountain made of silicon dioxide, the major constituent of many rock types such as quartz and granite. Beyond a certain height, the base of the mountain will melt out under its own weight and the mountain will start to collapse, so there must be a maximum height that is invulnerable to this instability. Using dimensional consistency, estimate the value of that maximum height.

. . . .

Our first task is to identify the collection of parameters that might influence the maximum height. Because the constraint on the mountain's height results from the melting of the solid silicon at the mountain's base, we can assume that the amount of energy needed to melt silicon will enter the picture. This quantity is called silicon's "heat of fusion." Heats of fusion, H, are usually expressed in units of joules/kg, and thus the fundamental units of H are energy/mass or $(\mathrm{ML}^2/\mathrm{T}^2)/\mathrm{M} = \mathrm{L}^2/\mathrm{T}^2$. Another parameter that might be relevant is the density of the rock, ρ, with units of M/L^3. Because the force of gravity is what causes the intense pressure at the mountain's base that can result in melting, the acceleration of gravity, g, with units of L/T^2, should be relevant. Next, we have the parameter of maximum height, h_{max}, itself, which has units of L. One other possible parameter in the problem is the cross-sectional area of the cylindrical mountain, A. A simple argument, however, suggests that this parameter should not influence our answer. The gravitational force on each segment of the mountain is directed straight downward toward the center of the Earth, and thus, when considering the effect of compression on the melting of rock at the base, you can think of the mountain as consisting of numerous vertical skinny columns of arbitrary area all bound together. For the base to melt, each of these columns has to melt separately and thus area cannot matter.

We seek a value for h_{max}, which has units of L, alone, so we need to find a combination of the other variables—H, ρ, and g—with units of L, alone. Writing this combination as $H^a\rho^b g^c$, we see the fundamental units are $(\mathrm{L}^2/\mathrm{T}^2)^{\mathrm{a}}(\mathrm{M}/\mathrm{L}^3)^{\mathrm{b}}(\mathrm{L}/\mathrm{T}^2)^{\mathrm{c}}$. Clearly we have to set $b = 0$, because otherwise the combination would contain mass units. To cancel the units of T, we must set $a = -c$, and hence the units of our expression are L^a. Therefore we require $a = 1$, and our unique, dimensionally consistent, expression for h_{max} becomes

$$h_{max} = \frac{cH}{g}, \tag{1}$$

where c is a dimensionless constant, which we presume is of order of magnitude 1.

What is the numerical value of this height? You can look up the heat of fusion of numerous substances in the *Handbook of chemistry and physics* (see Further Reading after the Appendix), where you will learn that for silicon dioxide in the form of quartz, H is 2.37×10^5 joules/kg. Using $g = 9.8$ m/sec^2, we then get

$$h_{max} \sim \frac{2.37 \times 10^5 \text{ joules/kg}}{9.8 \text{ m/sec}^2} = 24.2 \text{ km}. \tag{2}$$

Is this result a reasonable approximate answer to our initial question? It actually is. Had the answer turned out to be smaller than the heights of known mountains on the Earth's surface, that would have been either an embarrassment or a signal that the dimensionless constant, c, is really not of the order 1. When the same argument is applied to other planets, which of course have different values of the acceleration of gravity at the surface, the results are again consistent with known topography.

Better yet, a simple stability argument shows that h_{max} actually is equal to H/g. Imagine that a very thin disk at the base of a cylindrical mountain of height h melts, and let the thickness of the melted disk be δ. If the cross-sectional area of the mountain is A, the volume of melted mountain is δA, the mass of that much mountain is $\delta A\rho$, and the amount of heat needed to melt that much rock is $\delta A\rho H$. Now consider how much potential energy is lost when the mountain slumps by a distance δ. The mass of the mountain is dropping by a distance δ, and so the drop in potential energy is δ times the mass of the mountain times the acceleration of gravity. The mass of the mountain is $Ah\rho$, so the drop in potential energy is $\delta Ah\rho g$. This lost potential energy is converted to heat by the second law of thermodynamics, and so if we equate the release of this heat with the amount of heat needed to melt the disk, we get $h = H/g$. If h is less than this value, then the amount of potential energy that would be converted to heat if the mountain slumped down by an amount δ would not be enough to melt enough mountain to cause it to slump by an amount δ. If h is greater than H/g, then more than enough heat is released to cause the slump. Thus, H/g should be the maximum height of the mountain!

EXERCISE 1: Consider a cylindrical mountain made of ice instead of rock. How tall could it be? Does your answer surprise you?

EXERCISE 2: As a function of the ratio of a mountain's height to its diameter at the base, work out the maximum height of a more realistic, cone-shaped mountain made of silicon dioxide.

EXERCISE 3: Suppose a nuclear weapon explodes, creating a blast front that expands out from the point of detonation. If the speed of the expansion of this blast front is recorded photographically, write a dimensionally consistent expression for the energy released by the blast in terms of the data available from the time series of photos and from the known density of air. (This method was used to estimate the energy released by the atomic bomb tested at Alamagordo, New Mexico, in 1945.)

EXERCISE 4: How far from a 1-kiloton (4.18×10^{12} joules) explosion would you want to be to ensure that the moving blast front would not knock you down? (Make a reasonable assumption about what front velocity you could tolerate.)

III-2. Sleeping Bears

True hibernators (such as many rodents) metabolize during hibernation at a rate well below half their normal basal metabolic rate. When a typical 450-kg adult Alaskan brown bear (*Ursus arctos*) emerges from its winter hibernation, it is typically about 100 kg lighter than when it entered hibernation 6 months earlier. The fat it burned off yields about 38×10^6 joules/kilogram. Is this bear a true hibernator?

. . . .

To approach this problem, recall the statement earlier that the resting metabolic rate of adult mammals, from shrews to whales, varies as the three-fourths-power of body weight. Assuming that scaling relation, and using the fact that the basal metabolic rate of an adult human is 2×10^6 calories/day, you can estimate the basal metabolic rate of the Alaskan brown bear, which can then be compared with the energy yield from the fat burned by the bear during the winter.

The basal metabolic rate of a human being can be converted to the more convenient units of joules per second:

$$\begin{aligned}&\text{metabolic rate of human being}\\&\quad= \frac{2 \times 10^6 \text{ calories/day} \times 4.18 \text{ joules/calorie}}{(24)(3600) \text{ seconds/day}}\\&\quad= 96.7 \text{ joules/second.}\end{aligned} \tag{1}$$

A typical adult person has a mass of roughly 65 kg. Hence, using the metabolic scaling law, we can write

$$\begin{aligned}\frac{\text{metabolic rate of bear}}{\text{metabolic rate of a person}} &= \frac{(\text{mass of bear})^{3/4}}{(\text{mass of person})^{3/4}}\\&= \frac{450^{3/4}}{65^{3/4}} = 4.27.\end{aligned} \tag{2}$$

Hence, the resting metabolic rate of the bear is $\sim 4.27 \times 96.7$ joules/second = 413 joules/second. During the 6 months of hibernation, the bear consumes 100 kg of body fat. This amount is equivalent to $100 \times 38 \times 10^6$ joules. There are 3.15×10^7 seconds in a year (a number worth memorizing), and so 6 months of hibernation is equivalent to 1.58×10^7 seconds. Hence, the metabolic rate during those 6 winter

"Call it sleep?"

months is $100 \times 38 \times 10^6$ joules/1.58×10^7 seconds = 241 joules / second.
This rate is greater than half the estimated normal metabolic rate (413/2 = 207 joules / second), and so by the stated criterion, the bear is not a true hibernator!

EXERCISE 1: The basal metabolic rates of men and women differ somewhat; the former is about 6% greater than the latter. Moreover, men average about 20% greater weight than women. Is the difference between the metabolic rates of the two genders explainable by the three-fourths-power mass-scaling law? Comment on the statement, "the metabolic scaling law works remarkably well over a mass range from shrews to whales, so certainly it should work well over the much smaller mass range from women to men."

EXERCISE 2: A healthy adult person can (barely) survive a one-month food fast, provided water intake is maintained. What fraction of the body weight of an adult person and a 100-gram hamster will be lost during a one-month food fast if the person and the hamster are at rest the whole time? What are the implications of this for owners of small pets?

EXERCISE 3: Another pretty reliable scaling law for mammals asserts that lifetime scales as body weight to the one-quarter power. Given what you know about people, estimate the lifetimes of a 25-gram mouse, a 10-kg dog, and a 30,000-kg gray whale. Do your answers seem reasonable? What factors or assumptions could change your results?

EXERCISE 4: Based on the lifetime scaling law in Exercise 3, what is the scaling relationship that describes the dependence on body weight of cumulative basal metabolic activity over the lifetime of a mammal? Discuss the implications of your answer.

III-3. Little Green Men?

An old sci-fi story described people from a sunny climate who had evolved to have a symbiotic algal layer within their skin. By moving around in the sun all day, the people could live off the photosynthetic product of the algae. Is it physically possible for people to meet their food requirement this way? If not, how is it that some animals, such as certain marine copepods, do "farm" algae that grow on or within their bodies?

. . . .

Because the problem asks whether the situation is *physically* possible, we should look for some very fundamental constraint that might forbid the existence of such a lifestyle; it will not suffice to come up with arguments concerning the unlikelihood of the evolution of such creatures. Because "green animals" appear to exist if they are small but not if they are large, perhaps some scaling rule will allow us to understand the fundamental constraint. Let's make a simple "cylindrical cow" model of the situation and see if overall size affects the feasibility of living off a photosynthetically active surface.

Consider a cylindrical animal with a length of L meters and a radius of r meters. The volume of the animal is $\pi r^2 L$ cubic meters, and its mass is approximately the density of water times that volume. According to the metabolic scaling law, the metabolic requirement of that animal will be proportional to its mass to the three-fourths power, or $(r^2 L)^{3/4}$.

That metabolic rate must be sustained from the food eaten by the creature, and if the animal is to derive its energy from photosynthesis on its surface, the total rate of photosynthesis will be proportional to its surface area. That quantity is $2\pi rL$. Hence, the ratio of food supply to metabolic rate scales like $rL/(r^2 L)^{3/4} = L^{1/4}/r^{1/2}$. If that quantity becomes too small, then the animal cannot sustain itself with a surface-generated food source. If large and small animals had roughly similar body proportions, so that L/r remains roughly constant from shrews to people to whales, then the larger the animal the smaller the ratio of $L^{1/4}/r^{1/2}$. In fact, the ratio would scale like $1/L^{1/4}$. Thus, smaller creatures have a better chance than larger creatures of surviving through this lifestyle. Also, the longer and skinnier the creature, the more likely it can make a living this way, because, for a given volume, the surface area increases with increasing L/r.

This very general scaling argument tells us that smaller creatures, such as the copepod, are more likely than larger ones to be able to farm their surface area for a living. But it does not rule out the possibility for human beings nor endorse the possibility for copepods. To do that, we have to look at actual photosynthetic and metabolic rates.

An average person metabolizes at the rate of a lit 100-watt light bulb—in other words, about 100 joules/second. The most efficient algae known can convert solar energy into chemical energy at a rate of about 3% of the solar flux at their surface. The typical daily and annually averaged solar flux at the surface of the earth is about 170 watts/m^2, but in places it can be twice that high, and so food could conceivably be produced at a rate of about $0.03(340) \sim 10$ watts per square meter of surface area. Using the fact that the surface area of a typical person (see Appendix XV in *COW-1*) is about 1.7 m^2, we can then estimate that at the maximum feasible photosynthetic rate, people could supply about 17% of their basal metabolic requirement from algae growing all over their body. The actual percentage would be somewhat lower than this, however, because the algae need to use some of the chemical energy they produce for their own energy needs.

How small would a "little green man" shaped like a person (that is, with the same ratio of r to L) have to be so that it could gather 100% of its energy requirement this way? Such a creature would have to have a $100/17 = 6$ times greater ratio of food supply to energy requirement. If r is proportional to L, and the metabolic scaling law that works for mammals also holds for this creature, then the rule we derived above states that the ratio of food supply to energy requirement varies as $1/L^{1/4}$. To increase that ratio by a factor of 6, we need to decrease $L^{1/4}$ by a factor of 6, or L by a factor of $6^4 = 1{,}300$. People average about 1.5 m in length, and so the little green men have to be $\sim$0.001 m, or 1 millimeter, long to survive this way.

"And if you think these are incredible, you should see the REALLY big things on land here!"

Interestingly, that is about the length of the little copepods who exploit that food source! There are, of course, some issues we glossed over. In particular, copepods do not follow exactly the mammalian metabolic scaling law and they are not the same shape as people. In fact, they metabolize less than they would if they were a mammal with their body weight; on the other hand, they are rounder than people and thus have a smaller surface-to-volume ratio than would tiny, people-shaped organisms. These two corrections tend to cancel each other. Overall, our analysis illustrates the power of scaling laws to give us insight into wide-ranging phenomena.

EXERCISE 1: Let's take a naive approach to building a model to explain the mammalian scaling law relating body mass to metabolic rate. Mammals lose heat to the environment through their surface; thus, the rate at which body heat must be generated might be taken to be proportional to surface area. Body mass is proportional to body volume, however. So, consider spherical mammals—what would the metabolic scaling law be? Does this model explain the three-fourths-power scaling law? Now consider cylindrical mammals; how would the ratio of the length of the cylinder to its radius have to scale with body mass so that this surface-heat-loss approach would lead to the three-fourths-power scaling law? Is such dependence of shape (i.e., ratio of length to width) on body mass plausible? (The true origin of the three-fourths-power metabolic scaling law is actually much more interesting. It lies in the fractal space-filling property of the circulatory system in mammals. See West, G., Brown, J., Enquist, B. 1997. A general model for the origin of allometric scaling laws in biology. *Science* 276:122–26, for a fine article describing this law's origin.)

EXERCISE 2: Cities produce waste heat in proportion to the product of the number of people living in them and the per-capita rate of energy consumption. Cities shed waste heat to the atmosphere in two ways: by radiation, which should be proportional to area, and by wind advection, which should be proportional to the city diameter (see *COW-1*, Problem III.9). Both the radiative and advective losses will also be approximately proportional to the temperature difference between the urban area and the surrounding environment. Explore the implications of these statements for the way in which the intensity of the urban-heat-island effect (i.e., the temperature difference between the urban area and the environment) scales with population, with per-capita energy use, and with the shape of urban sprawl patterns with respect to the direction of prevailing winds. Think about such factors as how city area depends on population size and whether, for a fixed population size, city area will be fixed or somehow related to per-capita energy use. Will thermal impact, defined as the average temperature rise in the city, be simply proportional to the product of population size and per-capita energy use? Take a look at *COW-1* to get you started.

III-4. What a Drag

What is the ratio of the drag force caused by air resistance on a car moving at 15 m/second versus one moving at 30 m/second?

. . . .

This question may seem impossibly difficult to answer, given how little any of us knows about friction in moving cars, but let's plow ahead anyway in the spirit of the cylindrical cow. First, let's find a formula describing the dependence of the force of air resistance on relevant parameters. Just as we found the dependence of the period of a pendulum on its mass, length, and the acceleration of gravity, we will begin by trying to do the same for air resistance.

The relevant parameters you might plausibly come up with are the speed of the car, some measure of its area as "seen" by the air it encounters, and some property of the air itself.[14] The last of these needs further thought. If the car were moving through water, the resistance to motion would be far greater. How does water differ from air? For one thing, it is about 700 times denser (water has a density of 1,000 kg/m^3; air has a density at sealevel and 20°C of 1.3 kg/m^3). So, perhaps density is a relevant parameter.

Water also has a higher viscosity than air. Viscosity is a measure of how "sticky" a fluid medium is, that is, how much it resists internal motion. In Exercise 2, you will get a chance to explore the role of viscosity and how it affects air resistance at slow speeds. In this problem, with the automobile going at high speed through air, you will just have to accept for now that density and not viscosity is the relevant parameter characterizing the resistance the air offers to the moving object. Intuitively, it might make sense; after all, the car has to push big chunks of air out of the way as it barrels down the road. Also, the energy expended in moving that volume of air surely depends on the mass of the air, and thus, for some fixed volume of air, on its density.

In contrast, think about a tiny soot particle falling through the air. In that case, the mechanism of air resistance might well resemble the one that slows an object that is slightly denser than water falling slowly through water. In the latter case, the fluid is not so much shoved out of the way as it is parted. The density of the fluid medium is less important than is its "stickiness," and so under those conditions (sufficiently slow speed) viscosity is more important than density in determining air resistance.

---

14. The mass of the car should not matter because in the collisions between the car and the air molecules, the mass of the car is so much greater than that of the molecules that the molecules have, in effect, hit an infinitely unyielding wall.

How slow is slow enough for viscosity to be the governing quantity? To answer this question, scientists have identified an important dimensionless parameter called the Reynolds number. This parameter is explored in Exercise 4, where you will investigate its use in determining whether density or viscosity is the governing quantity. For now, let's return to our problem.

We seek an expression for the force exerted by air resistance against the moving car. The expression should be some function of the density of air and the effective area and speed of the car. The dimensions of these quantities, expressed as combinations of the fundamental dimensions M, L, and T, are M/L^3, L^2, and L/T, respectively. From them, we want to form an expression with dimensions of force. From the famous formula $F = ma$, we can deduce that the dimensions of force are ML/T^2. Hence, we must find coefficients a, b, c that satisfy $(M/L^3)^a (L^2)^b (L/T)^c = ML/T^2$.

Clearly, we must select $a = 1$ to match powers of M, and $c = 2$ to match powers of T. These choices then require that $b = 1$ to match powers of L. So, our dimensionally consistent combination of air density, area, and speed is

$$\text{air resistance} \propto (\text{density})(\text{area})(\text{speed})^2, \tag{1}$$

or in symbols,

$$F_{\text{air resistance}} = c\rho A V^2, \tag{2}$$

where c is a proportionality constant.

Isaac Newton was the first to formulate the relationship above for the resistance of a fluid to a fast-moving object, and the relationship is known as Newton's law of resistance. This law tells us immediately the answer to our problem: *doubling of car speed increases air resistance by a factor of 4.*

To know the absolute value of the wind drag on a car traveling at a given speed, you have to know the value of the constant, c, and the effective area, A, of the car. The former is undetermined by our application of dimensional consistency and must be determined by measurement. If we consider a spherical car of radius r, the effective area is πr^2, and the constant, c, is empirically very close to 1.[15] For typical cars, the product of the constant, c, and the effective area, A, is 3 or 4 m^2, somewhat greater than the actual frontal cross-sectional area because of the nonspherical shape.

---

15. This value is actually correct only for cars moving at ordinary speeds; as the speed of objects approaches Mach 1, the speed of sound, the constant becomes a function of the ratio of the speed of the car to the speed of sound, and the value rises with car speed. At Mach 1, the constant is about twice that at typical car speeds, but c varies only slightly between speeds of 5 and 50 m/second.

Thus, the actual air resistance at 30 m/second is given by

$$F_{\text{air resistance}} \sim (1.3)(3.5)(30)^2 \text{ kg m}^2/\text{second}^2 \sim 3{,}500 \text{ Newtons}, \quad (3)$$

where we have used the value $\rho = 1.3$ kg/m^3.

The total amount of work needed to overcome this air resistance during a drive of, say, 1 km is given by the equation: work = force × distance. Hence,

$$\begin{aligned} \text{work to overcome air resistance} &= (3{,}500 \text{ Newtons}) \times (1{,}000 \text{ meters}) \\ &= 3.5 \times 10^6 \text{ joules.} \end{aligned} \quad (4)$$

Using the fact that gasoline has an energy content of 4.8×10^7 joules/kg, you can compare this work with the total amount of energy needed to power a car. Exercise 1 gives you a chance to explore this comparison. You might want to consider two different situations: driving a car on level ground and driving up a hill; by so doing, you can compare the amount of energy needed to overcome gravity with that needed to overcome air resistance, and you can compare each of those amounts with the total energy needed to power the car.

EXERCISE 1: Roughly what fraction of the total energy required to power a car at 40 km/h, on level ground and up a hill, goes into overcoming air resistance? State your assumptions.

EXERCISE 2: Viscosity has dimensions of $M/(LT)$. The viscosities of air and water vary with temperature but at 0°C are 1.72×10^{-5} and 1.75×10^{-3} kg/meter-second, respectively. If the resistance acting against an object traveling at sufficiently slow speed through either medium depends on some combination of the object's speed, its cross-sectional area, and the viscosity of the medium, use dimensional consistency to determine the form of the relationship between the resistive force and the three dependent variables. In so doing, you will arrive at the form of Stokes' law for frictional drag. If you see the law as written in textbooks, you will find that the dependent variable "area of object" is replaced with "radius of object"; however, the dependence on area that you worked out above matches the function of radius that appears in the expression of the law.

EXERCISE 3: Your result in Exercise 2 determined the form of Stokes' law only up to an overall multiplicative dimensionless constant. Do a measurement to determine the value of the constant, using a spherical object so that effective radius can be taken to be the actual radius of the sphere. Drop a small, solid rubber ball (like the kind kids use to play jacks with) into a bucket of water, and from measurements of the time the ball takes to drop a known distance, work out the empirical value of the proportionality constant.

EXERCISE 4: Consider a sphere with radius r meters moving through a fluid medium at a speed of v meters/second. Ignoring the dimensionless constants in front of Stokes' law and Newton's law, construct a dimensionless quantity, R, which has the property that if R is smaller than 1, then viscous forces dominate and if R is greater than 1, then the inertial forces dominate. That parameter is called the Reynolds number, and it plays a big role in fluid dynamics.

EXERCISE 5: Evaluate R from Exercise 4 for a spherical droplet, 0.5 mm in diameter, falling from a cloud 2 km above the Earth's surface. [*Hint*: To evaluate v along the droplet's trajectory, you will have to solve $F = ma$, where F represents the combined forces of gravity and friction.]

III-5. Saving Species: The Role of Reserve Shape

Many factors enter into the design of nature reserves, and among them is the desire to include in the reserve as many species as possible for a given amount of preserved land. If the distribution of species is self-similar, would you expect long, skinny patches or squareish ones of the same total area to contain more species? More generally, how does the fraction of species in common to two censused patches depend upon the distance between the patches?

. . . .

Intuitively, the farther apart two patches lie in space, the more they should differ in the species they contain. The species near one end of a long, skinny patch of habitat are farther away in space from those at the other end than are the species inhabiting patches lying in opposite sides of a more square patch of the same area, so intuition suggests that long, skinny patches might contain the greater number of species. To convert this notion into a quantitative formula, we use the machinery of self-similarity as developed in the Background of this chapter.

The degree to which the collection of, say, plant species at one location differs from that at another can be quantified using a "commonality index," χ. χ is defined to be a fraction whose numerator is the number of species found in common at the two locations and whose denominator is the average of the numbers of species found at each of the two locations:

$$\chi = \frac{(S_1 \cap S_2)}{(S_1 + S_2)/2}, \tag{1}$$

where S_i is the number of species at site i, and $S_1 \cap S_2$ is the number of species in common to the two sites. Suppose that at each location, the species are censused in a patch of area A; for convenience, let the patch be a $\sqrt{2} \times 1$ rectangle as in the Background. Assuming self-similarity, we first identify the variables that χ might plausibly depend upon. Then we will use the fact that χ is a dimensionless fraction to write the most general dimensionally consistent expression for it. Finally we determine the actual functional form of χ.

The assumption of self-similarity implies that there is no fixed scale length in the problem—that is, no quantity with dimensions of area or length—other than the two spatial parameters that naturally appear in the problem: A, the area of the patches, and D, the distance between the centers of the two patches. Hence, we expect that χ depends only on A and D. Because χ is dimensionless, we have to form a dimensionless combination of A and D, the only one being A/D^2. Thus, we can assert that χ is some function of A/D^2. Without using further information, we can say nothing more specific, and so

$$\chi = \chi\left(\frac{A}{D^2}\right) \tag{2}$$

is the most general dimensionally consistent expression for the species commonality index. We would expect that commonality decreases with increasing distance between censused patches; thus, if the function above obeys a power law, then it goes like (A/D^2) to a positive, not a negative, power, but we cannot conclude anything more than Eq. 2 with dimensional arguments alone.

Now consider two censused patches within the landscape, one of area A with $S(A)$ species $[S(A) = cA^z]$ and one of area A' with $S(A')$ species. We assume, for now, that the distance D between the centers of the two patches is very large compared with either $(A)^{1/2}$ or $(A')^{1/2}$. We denote by $N(A, A', D)$ the number of species in common to the two patches. Commonality, defined as the number of species in common divided by the average number of species in the two patches, $[S(A) + S(A')]/2$, is then given by

$$\chi(A, A', D) = \frac{2N(A, A', D)}{S(A) + S(A')}. \tag{3}$$

Consider the probability that a species picked at random from the patch of area A' is also found in the patch of area A a distance D away. That probability, which we denote by $G(A, A', D)$, is given by the number of species in common to the two patches divided by the number in A', or from Eq. 3:

$$G(A, A', D) = \frac{N(A, A', D)}{S(A')}. \tag{4}$$

Using Eq. 3, we can eliminate $N(A, A', D)$ from this expression, to get

$$G(A, A', D) = \frac{[S(A) + S(A')]\chi(A, A', D)}{2S(A')}. \tag{5}$$

Now suppose we double the size of the patch of area A'. Each of the species in A' is still nearly exactly a distance D away from the center of the patch of area A [because $D \gg (A')^{1/2}$], and so if a species is selected at random from the patch of area $2A'$, the probability that it also lies in the patch of area A is unchanged. In other words, $G(A, A', D)$ is independent of A' and, in particular,

$$G(A, A, D) = G(A, 2A, D), \tag{6}$$

where A is an arbitrary area. Using Eq. 5,

$$G(A, 2A, D) = \frac{[S(A) + S(2A)]\chi(A, 2A, D)}{2S(2A)} \tag{7}$$

and

$$G(A, A, D) = \chi(A, A, D). \tag{8}$$

Eq. 6 tells us that the expressions in Eqs. 7 and 8 are equal, so

$$\chi(A, 2A, D) = \frac{2S(2A)\chi(A, A, D)}{S(A) + S(2A)}. \tag{9}$$

Now, using the species-area law in the form of Eq. 5 in this chapter's Background,

$$\chi(A, 2A, D) = \frac{2 \cdot 2^z}{(1^z + 2^z)}\chi(A, A, D). \tag{10}$$

Again considering $\chi(A, A', D)$, we can continue with this same reasoning. If we double the size of the patch of area A, then we increase the number of species in that patch and thus increase the probability that a species selected at random a distance D away from that patch of area A is also found there. The species number is increased by a factor of 2^z, and so the probability must increase by that same factor, or

$$G(2A, 2A, D) = 2^z G(A, 2A, D) = 2^z G(A, A, D). \tag{11}$$

Again using Eq. 5, $G(2A, 2A, D) = \chi(2A, 2A, D)$, and so

$$\chi(2A, 2A, D) = 2^z \chi(A, A, D). \tag{12}$$

Eq. 12 holds true when D and z remain constant and the censused areas double. Thus, Eq. 12 generalizes readily to the result $\chi(A, A, D) = (A)^z f(z, D)$, where f is some function. The form of the function f is constrained, however, because we saw that because of self-similarity and the fact that χ is a dimensionless fraction, $\chi = \chi(A/D^2)$. Therefore, we must have

$$\chi(A, A, D) = \left(\frac{A}{D^2}\right)^z g(z). \tag{13}$$

To determine the dimensionless function g of the dimensionless parameter z, we note that if the two patches share a common edge, so that $D^2 = A$ (recall that D is the distance between patch centers), then the number of species in common to the two patches, $N(A, A, A^{1/2})$, is exactly given by the difference between twice the number of species in a patch of area A and the number in the patch of area $2A$, or

$$N(A, A, A^{1/2}) = 2c(A)^z - c(2A)^z. \tag{14}$$

$\chi(A, A, A^{1/2})$ is given by dividing this expression for N by the average number of species in a patch, cA^z, or

$$\chi(A, A, A^{1/2}) = \frac{2c(A)^z - c(2A)^z}{c(A)^z} = 2 - 2^z. \tag{15}$$

But from Eq. 13, $\chi(A, A, A^{1/2})$ also equals $(A/A)^z g(z) = g(z)$, and hence we have

$$g(z) = 2 - 2^z. \tag{16}$$

Hence,

$$\chi \frac{A}{D^2} = (2 - 2^z)\left(\frac{A}{D^2}\right)^z. \tag{17}$$

Equation 17 is our desired formula for commonality in a self-similar world. The unsurprising implication of this formula is that patches farther away from each other are more different from each other with respect to the species they contain, so the intuitive statement that began III-5 is supported. But Eq. 17 provides a way to go beyond the qualitative statement and actually derive an explicit expression for the relative species richness of long, skinny patches versus $1 \times \sqrt{2}$ patches of the same area. Exercise 1 gives you an opportunity to do that.

This is truly an "all other things being equal" problem because many factors besides species richness affect the design of reserves. Among them are the desire to minimize the disturbances to the species inside the reserve that result from activities outside the reserve. Unfortunately, long, skinny reserves have a larger perimeter-to-interior ratio than do $1 \times \sqrt{2}$ patches of the same total area; more of the interior of a long, skinny patch is near a border. Thus, the species in long, skinny reserves are likely to suffer more disturbance. As in so many situations, tradeoffs must be made.

EXERCISE 1: By repeated use of the commonality formula, Eq. 17, derive an approximate formula, in terms of only the parameter z, for the ratio of the number of species in the following two rectangles

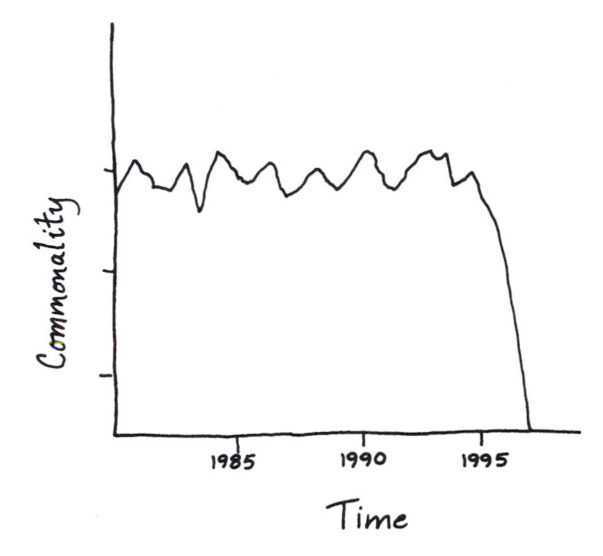

"We suspect the precipitious decline in floral commonality between plots A and B has something to do with the shopping mall built on top of plot B in 1996. Next slide."

of the same area:

$$\text{rectangle 1:}\quad 4 \times 4\sqrt{2}$$
$$\text{rectangle 2:}\quad 1 \times 16\sqrt{2}.$$

The formula will be approximate because you will be repeatedly assuming Eq. 17 describes the fraction of species in common to two long, skinny rectangles that share a common short edge. Evaluate the ratio for the specific case of $z = 0.25$. What can you conclude about the implications of self-similiarity for the dependence of species richness on the shape of censused patches?

EXERCISE 2: As mentioned in the last paragraph of Problem III-5, a conflict exists between the desire to maximize the number of species found, at least initially, in a protected area and the desire to minimize the disturbances to these species that arise from activities outside the reserve. To find an optimum length-to-width ratio for a preserve, some explicit assumption about disturbance must be made so that the consequences of disturbance can be converted into quantitative terms that can be balanced against the quantitative shape factor, which Eq. 17 expresses. Make a reasonable such assumption about disturbance, and then carry out the optimization calculation.

Chapter IV
Core Models of Change in Time and Space

Gonzalo: "You are gentlemen of brave mettle; you would lift the moon out of her sphere if she would continue in it five weeks without changing."

Introduction

The differential calculus was coinvented by Isaac Newton to solve a problem that, at the time, was incredibly difficult: What is the trajectory of objects like moons and apples whose movement is governed by Earth's gravitational force? Thanks to Newton's invention, bright undergraduates can now solve the problem. In the study of climate, hydrology, ecology, and virtually every other area of environmental science, differential calculus is an essential tool today. Figuratively speaking, Newton provided us with a way to take things apart (differential calculus) and put them back together again (integral calculus). Why is this so important? The reason is that scientific laws and assumptions are usually cast in terms of rules about instantaneous interactions among things or in terms of descriptions of infinitesimal elements of macroscopic objects. Yet what we seek is information about macroscopic behavior, shape, and motion. The calculus provides, among other things, a way to convert such laws and interactions into the desired output. In this section, you will become acquainted, and I hope comfortable, with a small but useful core repertoire of calculus-based models for studying the environment.

Background

The models we explore in this chapter are all described by differential equations. These models are constructed to permit study of how the components of a system change over time and space. The differential equations relate the rates of change over time and/or the degrees of variation over space in the components to the actual values of the components. Before we plunge into an examination of the ways actual models are constructed and solved, let's get familiar with the taxonomy of differential equations and the language used to describe the many varieties. The eight dichotomous choices listed below provide such a taxonomy.

1. linear vs. nonlinear
2. homogeneous vs. inhomogeneous
3. first order vs. higher order
4. single variable vs. coupled multivariable
5. time-independent vs. time-dependent
6. ordinary vs. partial
7. instantaneous vs. time-lagged
8. deterministic vs. stochastic

To understand the meaning of these terms, consider the venerable differential equation

$$\frac{dX(t)}{dt} = rX(t).$$

The left side of this equation is the rate of change of X with respect to time. The right side tells us that this rate of change of X is equal to a constant, r, times X itself. We refer to this as the Malthus equation because it is the differential equation that describes the unchecked growth over time, t, of a population, $X(t)$, under certain conditions that Malthus described in his famous 1798 treatise. The term r is a constant, which is determined by the birth and death rates for the population. This equation is about as simple a differential equation as we can write. It is characterized by the simpler of the two choices in each of the eight categories in the list above, which means the left choice in each case. Thus, the Malthus equation is a linear, homogeneous, first-order, single-variable, time-independent, deterministic, instantaneous, ordinary differential equation. Indeed, it is the only such equation!

To explain, the Malthus equation is

- **linear** because X appears only to the first power; if terms such as X^2 or $\log(X)$ appeared in the equation, then it would be **nonlinear**.
- **homogeneous** because each term in the equation is proportional to X; if the equation had read $dX/dt = rX + b$, where b is a constant, then it would be **inhomogeneous**.
- a **first-order** differential equation because only the first derivative of X with respect to t appears; if the equation involved the second derivative, d^2X/dt^2, or higher derivatives, then it would be a **higher-order** differential equation.

- a **single-variable** differential equation because only one system variable, X, appears in it; if we dealt with two or more interacting populations, such that the rates of change of each depended upon the others, then we would need to write a **coupled multivariable** equation.
- **time-independent** because time appears in only two places: as the variable upon which X depends and explicitly in the time derivative; if r were an explicit function of time, because birth or death rates were changing over time, then the equation would be **time-dependent**.
- an **ordinary** differential equation because a derivative is taken with respect to only one variable, t; if X depended on space as well as time, and the differential equation describing X involved derivatives with respect to space as well as time, it would be a **partial** differential equation (or PDE, as such equations are affectionately called by their intimate friends).
- an **instantaneous** differential equation because dX/dt and X are evaluated at the same time, t; if the rate of change of X at time t depended upon the history of X prior to time t, then it would be a **time-lagged** differential equation.
- **deterministic** because the only parameter in the equation, r, is a true constant; if r varied randomly over time, it would be a **stochastic** equation. Note that a stochastic equation must be a time-dependent equation, so the choice here really comes into play only if the equation is time-dependent.

Do you now see why the Malthus equation is the only equation described by all the left-side terms in the taxonomic list?

Rewriting the equation in the form $dX/X = r\,dt$, and using the fact that $\int dX/X = \ln(X) + \text{constant}$ and that $\int r\,dt = rt + \text{constant}$, we can solve the equation by integration:

$$\ln(X) = rt + c,$$

where c is an overall constant of integration. This equation can be rewritten as

$$X(t) = ce^{rt}.$$

If we knew the value of X at some particular time t_0, then we could determine the as-yet undetermined constant c. In particular, knowing the numerical value for $X(t_0)$, we can set

$$c = X(t_0)e^{-rt_0}.$$

Often, t_0 is chosen to be time 0; in that case, since $e^0 = 1$, $c = X(0)$.

Because the Malthus equation is first order and ordinary, there is only one constant of integration, c, and it can be determined by one piece of knowledge: the value of X at some time t. If the equation involved the second-time derivative of X, then we would require two pieces of information to determine the two constants of integration that would arise. These data might, for example, be the values of X at two times, or the value of X and of its first derivative at one time.

The Malthus equation describes exponential growth if r is positive. The constant r is the rate constant describing that growth, and if X is the size of a population, then r is the difference between the per-capita birth rate and per-capita death rate. If X is the amount of money in an interest-bearing bank account, then r is the interest rate compounded over a time interval equal to whatever the time unit is in which the parameter t is expressed.

The Malthus equation also describes exponential decay if r is negative. The decay might be that of a collection of radioactive atoms, in which case the magnitude of r is the decay rate constant characterizing the particular species of atom. The decay might also be that of a population sliding to extinction because the death rate exceeds the birth rate.

Exponential growth cannot continue forever, of course, so as a model of the change over time in a growing population, the Malthus equation must be incomplete. Some of the problems that follow provide you with insight into ways to modify the model to make it more realistic in an ecological context. You will also explore relatively simple models that are of value in understanding climate and other phenomena in nature. All of these models will be more complex than the simple Malthus model and thus will contain some of the right-side characteristics from the list of taxonomic characteristics of differential equations.

My approach in this chapter is to familiarize you with the differential equation models that lie at the core of each of several fields of environmental science. Each of the core models is a starting point for more advanced modeling approaches, but often the insights derived from more complex models actually stem from the simpler models that lie at the core. Some of the exercises that follow these problems give you an opportunity to explore more advanced versions of the core models.

Problems

IV-1. Pollutant Stock-and-Flow: Reconstructing CO_2 Emissions

At the start of the industrial revolution, around 200 years ago, the concentration of CO_2 in the atmosphere was about 275 ppm(v); 100 years ago (1900) it was about 285 ppm(v); and currently (2000) it is about 365 ppm(v). Emissions from fossil-fuel burning have been growing roughly exponentially over the past 100 years and today are about 5.5×10^9 tons(C)/y, or, recalling that the prefix *giga* means 10^9, 5.5 Gt(C)/y. Another ~2 Gt(C)/y are estimated to be emitted from burning or decomposition of deforested biomass, resulting in a total of about 7.5 Gt(C)/y emitted. The net increase in the amount of CO_2 in the atmosphere is currently averaging about 2.5 Gt(C)/y. The difference, $7.5 - 2.5 = 5$ Gt(C)/y, is the current net rate at which the oceans and the land must be taking up the remaining portion of annual atmospheric CO_2 emissions.

Let's see what we can learn, using a simple model, about the history of fossil-fuel consumption and the mechanisms of excess CO_2 removal. Assume, first, that the net rate of removal of CO_2 from the atmosphere during this time has been proportional to the difference between the actual level and the preindustrial level, and that the deforestation rate has been relatively constant over the past 100 years. What has been the doubling time for CO_2 emissions from fossil-fuel burning over the past 100 years? You will find that the answer to this question is unreasonable, and thus at least one of our assumptions must be wrong. So, the second question is: From the failure of our simple model, what can be learned about the flows of carbon in and out of the atmosphere?

. . . .

Consider, first, the buildup of pollution in a lake. Assume that pollution enters the lake at a constant rate, F_{in}, and pollution exits the lake at a rate proportional to the amount of pollutant in the lake.[16] From these assumptions, the following linear, donor-controlled differential equation describes the time-dependent amount of pollution in the lake, $P(t)$:

$$\frac{dP(t)}{dt} = F_{in} - kP(t). \tag{1}$$

As we showed in *COW-1*, the proportionality constant, k, equals the inverse of the hydraulic residence time[17] of water in the lake if the only mechanism of pollutant removal is outflow from the lake in the outgoing stream water. The general solution to Eq. 1, subject to the initial condition $P(t = 0) = P_0$, is

$$P(t) = \left(P_0 - \frac{F_{in}}{k}\right)e^{-kt} + \frac{F_{in}}{k}. \tag{2}$$

Note that this general solution satisfies the initial condition and that as $t \to \infty$, $P \to F_{in}/k$.

Eq. 1 is the basic core model used in the analysis of stocks and flows of nonliving entities (e.g., pollutants). If the entity is alive, then the assumption of constant input is no longer valid (see Problems IV-2, -4, -5). In terms of our taxonomy of differential equations describing system dynamics, Eq. 1 differs from the Malthus equation only in that it is inhomogeneous. Our CO_2 problem requires a generalization of this equation to the case in which the input is not constant in time, so the equation we construct will be time-dependent. It provides a useful review of how simple box models are formulated and used to try to answer important questions.

As with all input–output box models, the first step in constructing a model is being clear about the system boundaries and components, the major stocks and flows, and the units with which we will describe the stocks and flows. In our case, there is a single box—the entire atmosphere. Confusion about the substance we are describing can arise be-

16. The first assumption is reasonable if the discharge of pollution to the lake occurs at a constant rate. The second is reasonable if the pollutant is well mixed within the lakewater and exits the lake dissolved in a constant, flowing stream of water from the lake. If the outgoing stream flow is not constant but instead varies during the year, then the second model assumption requires that the variables refer to annual averages. Of course, if there is interannual variation in the outflow rate, the assumption makes sense only if even longer time averages are taken.

17. Hydraulic residence time is the stock of water in the lake divided by the rate of stream outflow. If water exits the lake by evaporation as well as stream outflow, then the residence time of water (stock divided by total outflow rate) will be less than the hydraulic residence time.

cause the problem statement talks about both carbon dioxide and carbon. To be consistent, we will always take our stock variable to be carbon, C, rather than CO_2. For our stock, we will use units of Gt(C) and for the flows we will use Gt(C)/y. Thus, our time unit is years. The atmospheric concentrations in the problem statement are in units of parts per million by volume (or equivalently by moles), which can be converted to Gt(C) by the following conversion (Exercise 1):

$$1 \text{ part per million by volume of } CO_2 = 2.16 \text{ Gt(C)}. \quad (3)$$

The second step is determining the functional dependence of the flows upon the stocks and explicitly upon time. In our case, the input rate from fossil-fuel burning is assumed to grow exponentially, but the growth rate constant is not specified (in fact, that constant is what we have to determine). So let's write

$$F_{\text{in, fossil fuel}} = ae^{bt}, \quad (4)$$

where a and b will have to be determined. The other input process is deforestation, which we are told to assume has been constant:

$$F_{\text{in, deforestation}} = 2, \quad (5)$$

where we leave off the units because we have agreed what they are. Together, this gives us the total input rate:

$$F_{\text{in}} = ae^{bt} + 2. \quad (6)$$

We are told to assume that at any time t, the output rate is proportional to the difference between the concentration of carbon in the atmosphere at that time, $C(t)$, and the preindustrial level of 277 ppm(v), or 598 Gt(C).[18] Hence,

$$F_{\text{out}} = k[C(t) - 598], \quad (7)$$

where k is the proportionality constant.

The third step is constructing a differential equation by setting the rate of change of the stock to the input rate minus the output rate:

$$\frac{dC}{dt} = ae^{bt} + 2 - k[C(t) - 598]. \quad (8)$$

18. This assumption is supported by the plausible idea that atmospheric carbon dioxide in excess of the equilibrium level prior to the era of fossil-fuel burning will tend to be transferred to Earth's surface, so as at least partially to restore the equilibrium; this concept is similar to the familiar observation that a dissolved substance initially concentrated in one end of a lake will tend to disperse to the less polluted waters. Numerous measurements, as well as atomic theory, confirm the assumption that the restoring flux is linearly proportional to the concentration difference.

Before proceeding to solve this equation, let's see how much we already know about the parameters a, b, and k. From the fact that the net increase in the atmosphere is currently 2.5 Gt(C)/y, we have a constraint on the parameters. Let's take time $t = 0$ to be 100 years ago. Time today, then, is $t = 100$. The current value of C is given, and the net annual increase in C is given by $dC/dt \times 1$ because time is measured in units of years. So evaluating dC/dt at current time, and using $C(100) = 2.16 \times 365 = 788$ Gt(C), we get one relationship among a, b, and k:

$$ae^{100b} + 2 - k(788 - 598) = 2.5. \tag{9}$$

We also know that the current rate of emission is 5.5 Gt(C), or:

$$ae^{100b} = 5.5. \tag{10}$$

Combining these two equations, we have

$$k(788 - 598) = 5, \tag{11}$$

or

$$k = \frac{5}{190} = 0.0263/\text{y}. \tag{12}$$

To proceed further, we have to solve Eq. 8. This differential equation is an example of a general class of differential equations of the form

$$\frac{dX(t)}{dt} = \alpha(t)X(t) + \beta(t). \tag{13}$$

Its most general solution[19] is

$$X(t) = e^{\int \alpha(t')\,dt'}\left\{c + \int \beta(t')e^{-\int \alpha(t'')\,dt''}\,dt'\right\}. \tag{14}$$

Here, the expression $\int f(t')\,dt'$ means "the indefinite integral of the function f evaluated at t," and the expression $\int f(t'')\,dt''$ means "the indefinite integral of the function f evaluated at t'." The parameter c is the constant of integration.

In our equation, $\alpha(t) = -k$ and $\beta(t) = 598k + ae^{bt} + 2$. Using

$$\int dt' = t, \tag{15}$$

19. Further detail on this solution and the solutions to many other classes of differential equations applicable to environmental science can be found in any of the several excellent textbooks on differential equations; see, for example, Boyce, W., DiPrima, R. 1986. *Elementary differential equations*, 4th ed. New York: Wiley.

and

$$\int e^{kt'}\,dt' = \frac{e^{kt}}{k}, \tag{16}$$

and

$$\int e^{bt}e^{kt'}\,dt' = \frac{1}{b+k}e^{(b+k)t}, \tag{17}$$

we get

$$C(t) = ce^{-kt} + 598 + \frac{2}{k} + \frac{a}{b+k}e^{bt}. \tag{18}$$

Now we can impose the additional information that $C = 2.16 \times 285 = 615.6$ Gt(C), which we round to 616 Gt(C) at time $t = 0$ and $C = 788$ Gt(C) at time $t = 100$:

$$C(0) = c + 598 + \frac{2}{k} + \frac{a}{b+k} = 616 \tag{19}$$

and

$$C(100) = ce^{-100k} + 598 + \frac{2}{k} + \frac{a}{b+k}e^{100b} = 788. \tag{20}$$

If we now use the value of k given by Eq. 12, we can solve Eqs. 9, 19, and 20 for the three remaining unknowns: a, b, and c. The three equations are sufficiently complicated that a numerical solution is required. This yields

$$a \cong 0.74 \text{ Gt(C)/year}, \tag{21}$$

$$b \cong 0.020/\text{year}, \tag{22}$$

and

$$c \cong -74 \text{ Gt(C)}. \tag{23}$$

The doubling time for CO_2 emissions can be determined from the rate constant, b. Recall that the ln of 2 is 0.693. Hence, $e^{bt} = 2^{bt/0.693} = 2^{t/T_2}$. Here, T_2 is the doubling time, and it is easily seen to be given by

$$T_2 = \frac{0.693}{b}. \tag{24}$$

Substituting in the value from Eq. 22, we arrive at a doubling time of 35 years.

Unfortunately, this answer is not correct. The global rate of fossil-fuel consumption has actually been doubling roughly every 20 years during the past 100 years, and thus it has been growing at a rate of $b \sim 0.035/\text{y}$, nearly twice our result in Eq. 22. Our solution predicts that 100 years ago, global fossil-fuel consumption led to the emission of about 3/4 Gt(C)/year, whereas in reality the rate then was about 1/6 Gt(C)/y.

So something must be wrong with our assumptions. One of our assumptions was that carbon dioxide emissions resulting from deforestation occurred at a constant rate of 2 Gt(C)/y throughout the century. This actually overstates historic deforestation, but if we corrected this error, our answer would be even more incorrect! The reason is that a lower historic deforestation rate requires a larger value for fossil fuel burning in the year 1900 and a lower exponential growth rate constant.

Perhaps our use of the baseline value of 598 Gt(C) in Eq. 7 is incorrect, and the sink strength at least during this past 100 years is proportional to the difference between $C(t)$ and some other baseline level C_0:

$$F_{\text{out}} = k[C(t) - C_0]. \tag{25}$$

So let's look at the solution to the equation

$$\frac{dC(t)}{dt} = ae^{bt} + 2 - k[C(t) - C_0], \tag{26}$$

taking $b = 0.035$, and solving for C_0.

We proceed as before, solving Eqs. 9, 10, 19, and 20, but with b set to the value of 0.035 and with the numerical value 598 replaced by the unknown C_0. This step yields the numerical solutions

$$k \cong 0.044/\text{y}, \tag{27}$$

$$a \cong 0.17\ \text{Gt(C)/y}, \tag{28}$$

$$c \cong -106\ \text{Gt(C)}, \tag{29}$$

$$C_0 \cong 674\ \text{Gt(C) or 312 ppm(v)}. \tag{30}$$

This solution solves one problem, for it enforces the correct value of b, but it raises another problem. If Eqs. 26 through 30 are taken at face value, then in the years prior to 1956, when the carbon dioxide level in the atmosphere reached ~674 Gt(C), the term $k[C(t) - C_0]$ in Eq. 26 was actually negative. In other words, for the first half of the century, the Earth's surface would have been a net *source* of carbon dioxide to the atmosphere if deforestation and fossil-fuel burning are neglected. This result appears peculiar; the level of carbon dioxide in the atmo-

"He's been sniffing CO_2 again."

sphere during that period was well above the preindustrial level, suggesting that the $k[C(t) - C_0]$ term should correspond to a net flux from the atmosphere back to the surface.

There are a few flows of carbon dioxide to the atmosphere that we have left out of the model, such as a small source from cement manufacturing. Moreover, the growth of fossil-fuel use has not been strictly exponential. These corrections are not even close to being sufficient to solve the problem, however. Another, more likely, possibility is that the excess carbon dioxide has fertilized the forests of the world, increasing the rate of photosynthesis and leading to a net positive change in the balance between photosynthesis and decomposition. Trees take time to grow, but as they grow bigger, they can incorporate even more carbon dioxide, so this effect might increase disproportionately faster than the increase in atmospheric carbon dioxide and thus might show up more strongly in the second half of the century. Exercise 7 gives you a chance to explore this idea.

EXERCISE 1: Verify Eq. 3.

EXERCISE 2: Show by direct substitution that Eq. 2 satisfies Eq. 1 and that Eq. 14 satsifies Eq. 13.

EXERCISE 3: If the rate of exponential growth in energy use, $b = 0.028$, that we used in deriving Eqs. 27 through 30 were applicable to the time period prior to 100 years ago, what would have been the annual rate of fossil-fuel-caused carbon emissions 200 years ago, at the nominal beginning of the industrial revolution? Is it plausible that a constant rate, b, has actually characterized emissions reasonably well over the entire 200-year period?

EXERCISE 4: Assuming that our Eqs. 26 through 30, with $b = 0.028$, characterize the future carbon dioxide budget, in what year will the CO_2 concentration in the atmosphere reach 700 ppm(v)?

EXERCISE 5: Our two proposed candidates for resolving the puzzle raised by Eq. 30 are (a) deforestation and (b) stimulation of forest growth caused by the fertilizing effect of increased carbon dioxide in the atmosphere. These factors appear quite opposite to each other! How can they both explain the same thing?

EXERCISE 6: If fossil-fuel emissions were to cease suddenly (for example, if the international climate-change negotiations to reduce fossil-fuel consumption were to succeed beyond my wildest hopes), then Eq. 26 predicts that the CO_2 level in the atmosphere would decline over time to the value C_0. Given the numerical solutions in Eqs. 27 through 30, we see that the level would never return to the preindustrial value of 275 ppm(v)—we would be left with 312 ppm(v). Indeed, a cessation of anthropogenic emissions would not result in a recovery back to 275 ppm(v), not even in 1,000 years, but the reason has to do with a mechanism not included in our simple model—the carbonate-buffering effect in seawater. *COW-1* (pp. 138–49) discusses that mechanism in detail and shows how to calculate the Revelle factor, which determines the percentage of anthropogenic carbon dioxide that will remain in the atmosphere for millennial time scales. To delve further into carbon sinks, read that material in *COW-1*, and then try to construct and analyze a more realistic model for carbon dioxide removal that includes this carbonate-buffering effect.

EXERCISE 7: Delving still further into the mechanisms for atmospheric carbon removal, suppose that the historic increase in the atmospheric CO_2 level has fertilized the earth's forests, leading to increased uptake of atmospheric carbon dioxide by trees. Assume, further, that increased tree biomass results in an increased rate of uptake of CO_2. To study the consequences of these assumptions, construct a two-box model involving both forest biomass and the atmosphere, and use it to explore the possibility that forests comprise a carbon sink of increasing strength. You may want to wait until you have finished reading Problems IV-4 and IV-5 before you tackle this one, unless you already have some idea of how to construct models that incorporate the dependence of the growth rate of an organism on its food supply. To be even more ambitious, include in your model the organic carbon stored in soils that will eventually be decomposed to CO_2. For this portion, you will almost certainly want to wait until you have read Problem IV-5 on biogeochemistry.

EXERCISE 8: The following are reasonable assumptions about several important carbon stocks and flows in recent years: (a) Of the total anthropogenic emissions of carbon dioxide to the atmosphere [~7.5 Gt(C)/y], about 80% originates in the northern hemisphere and

20% in the southern. (b) The interhemispheric mixing time for atmospheric CO_2 is 1 year (as derived on pp. 40–44 in *COW-1*). (c) The northern hemispheric atmosphere currently contains about 2 Gt(C) more CO_2 than the southern hemispheric atmosphere. (d) The rate of buildup of CO_2 in the atmosphere is ~2.4 Gt(C)/year, and the increase is nearly equal in both hemispheres. For these assumptions to be consistent with each other, what must be the ratio of the rate of downward removal of CO_2 from the northern hemisphere's atmosphere to the corresponding rate in the southern hemisphere? Under the assumption that the downward removal rates to land and sea are proportional to the areas of these two surface categories, does your answer above suggest a land or sea sink for CO_2? Why? What neglected factors or information might render misleading this method of determining whether the CO_2 sink is mainly land or sea?

IV-2. Population: Limits to Growth

Imagine a single bacterium in a bottle containing nothing but air and bacterial growth medium. The air in the bottle is exchanged with outside air at a fixed rate in an attempt to control the level of toxic gases such as ammonia, methane, or carbon dioxide that are steadily produced by the live organisms. Although the rate at which the bacterial cells divide is a fixed constant, their death rate is proportional to the concentration of toxic gas in the bottle. Additional growth medium is added whenever needed to ensure that the bacteria are never food-limited, but as their numbers increase and the population grows more crowded, they are increasingly poisoned by their own waste products. Construct a differential equation that describes the change over time in the population, and contrast the solution to that equation with the solution to the Malthus equation.

. . . .

The starting point for any population growth problem is a Malthus-type equation in which birth and death processes are characterized. Let's call $N(t)$ the number of bacteria in the bottle at time t, and let's take the initial time, when there is a single cell in the bottle, to be $t = 0$, so that $N(0) = 1$. Very generally, the differential equation governing N is

$$\frac{dN(t)}{dt} = (B - D)N(t), \tag{1}$$

where B is the per-capita rate of cell division (that is, the total number of cell divisions per unit time divided by the number of cells) and D is the per-capita rate of cell death. The distinction between the per-capita birth or death rates and the total rates is important to keep clear. The total death rate is $DN(t)$, whereas D is the death rate per cell. D has units of $(\text{time})^{-1}$. Expressed in units of, for example, $1/\text{day}$, D is the fraction of cells that will die in any day; multiplied by N, it gives the actual number of cells that die in a day.

According to the problem statement, the rate at which cells divide, or B, is a fixed constant. But D is not a constant; it is proportional to the amount of toxic gas in the bottle, which will change as N changes. Hence,

$$D = c_x X(t), \tag{2}$$

where $X(t)$ is the amount of toxic gas in the bottle and c_x is a proportionality constant. Combining Eqs. 1 and 2, we have $dN/dt = BN - c_x XN$. So, before we can solve the problem and determine N, we have to make a model for the amount of toxic gas, X, in the bottle.

Because living bacteria produce the toxic gas, the rate of input of toxic gas to the bottle is proportional to N. The rate of output is going to be governed, in part, by the rate of air exchange between the bottle and the outside air. It will also depend upon the concentration of toxic gas in the bottle. We encountered a situation something like this before, in *COW-1*, when we looked at the buildup of pollution in a lake. There, we had a constant rate of input of pollution to the lake and an output of pollution because of water exiting in an outflow stream. As a result of the fixed rate of water outflow, pollution exited from the lake at a rate that was proportional to the concentration of pollution in the lake. This information led us to a differential equation for the amount of pollution, P, in the lake. The equation was $dP/dt = F_i - aP$, where F_i was the rate of pollution input to the lake and the parameter a was proportional to the rate at which water exited the lake in the outflow stream. If you start with a clean lake, we saw that this equation tells us that the pollution builds up and approaches an ultimate steady state given by $P = F_i/a$. The concentration of pollution in the lake would be just P/W, where W is the constant amount of water in the lake.

Again using X to denote the amount of toxic gas in the bottle, then

$$\frac{dX}{dt} = F_i - F_o, \tag{3}$$

where F_i is the rate of production of toxic gas by the bacteria and F_o is the rate of removal by air exchange. In our lake problem, we had a constant-rate pollution input to the lake from a nearby factory that operated at a constant rate. In our present problem, F_i is proportional to $N(t)$. So, let $F_i = c_i N(t)$, where c_i is some constant characterizing bacterial metabolism. Just as in the lake problem, the output rate, F_o, is proportional to the amount of X in the bottle because a fixed fraction of the air in the bottle is removed in every unit time interval. Hence, we set $F_o = c_o X$, where c_o is some constant characterizing the rate of air exchange, and we arrive at a differential equation for X:

$$\frac{dX(t)}{dt} = c_i N(t) - c_o X(t). \tag{4}$$

This equation must somehow be combined with our differential equation for N,

$$\frac{dN(t)}{dt} = BN(t) - c_x X(t)N(t), \tag{5}$$

and the coupled, multivariable system of equations then solved for $N(t)$. The equations could be solved numerically, and in an exercise you will have a chance to do that. For now, we make a reasonable approximation that will readily allow us to obtain a famous, single-variable differential equation in ecology that can then be solved analytically to obtain an exact solution. We assume that the rate of change of X, dX/dt, is very small compared with either term on the right side of Eq. 4. In other words, $c_i N \sim c_o X \gg dX/dt$. If that assumption is valid (you will explore the circumstances under which it is valid in Exercise 6 below), then we can approximate X by $X \sim c_i N/c_0$. Substituting this value for X into Eq. 5, we arrive at the differential equation

$$\frac{dN}{dt} = BN - \frac{c_x c_i}{c_o} N^2. \tag{6}$$

If we set $B = r$ and $Bc_o/c_x c_i = K$, then Eq. 6 can be rewritten in the form

$$\frac{dN}{dt} = rN\left(1 - \frac{N}{K}\right). \tag{7}$$

This is the famous **logistic equation**, the mother of (nearly) all ecological models.

What does the logistic equation say, and why is it so important? You will get to actually solve the equation in Exercise 1, below, but let's see how much we can learn about the solution just by staring at the equation. If N starts out very small, as is the case in our bottle, with only 1 bacterium initially, then it is likely that $N/K \ll 1$, so that the equation is approximately the Malthus equation, $dN/dt = rN$. We know what the solution to the Malthus equation looks like—exponential growth. But as N grows, eventually, N/K will not be far less than 1, and the effect on the growth rate of the $1 - N/K$ term has to be considered. As long as $N/K < 1$, the term is positive, and so the overall sign of the right side of Eq. 7 will be positive. That means growth continues, but the rate of growth decreases as N approaches K. When $N = K$, growth ceases because the right side is then zero. We call that a steady-state solution to the equation. So, $N = K$ is a steady state, and once $N = K$, the population size will remain equal to K forever.

We deduce from this consideration that the shape of a curve of $N(t)$ versus t will start out looking like exponential growth, but as N increases, the rate of growth declines, and N eventually levels out and approaches a steady-state value given by K. K is called the bacterial carrying capacity of the bottle. We note that from its definition, K is proportional to c_o and inversely proportional to c_i. The former means that if the air-exchange rate in the bottle were higher, or the bottle were bigger, then the carrying capacity would be greater; the latter means that if bacteria emitted less waste product, the carrying capacity would be greater.

As a model of a growing population, the logistic equation is often described as incorporating "density dependence." This term refers to the fact that the total death rate in the logistic model is not a constant times N but rather a constant times N^2. Hence, the per-capita death rate is not a constant as it is in the Malthus equation, but rather it is proportional to N itself. Because the density of the population affects the per-capita death rate, we say that the logistic equation expresses density dependence. Density dependence is ubiquitous in nature, not only in bottled populations of a single species but in real ecosystems with interacting species. It can take many forms and can arise from many underlying mechanisms besides the self-poisoning effect considered here (see, for example, Exercise 7). It may affect birth as well as death rates, and predation rates as well. Density dependence has an enormous effect not only on curbing unlimited growth but also, as we will see in Chapter 5, on the stability of ecosystems.

The logistic equation is at the core of (nearly) all ecological models in the sense that buried within most such models are logistic equations. Like the complex tangle of technological gadgetry that Dorothy encountered in Oz, at the center of which was just a "wizard" pulling a few ropes, many ecological models adorned with bells and whistles have at their operating centers merely our logistic equation. Density dependence need not always be expressed in exactly the form of Eq. 7, but the idea behind Eq. 7 pervades much of population and ecosystem modeling. Interestingly, if the experiment described by the problem statement is carried out, the growth of the microbial population is generally described rather well by the solution to Eq. 7.

EXERCISE 1: Classify the logistic equation according to the taxonomic key to differential equations presented in the Background section.

EXERCISE 2: Consider the town or city in which you live and the causes of human death there. Based on actual reasons people die there, explain why you think the human death rate is or is not significantly density-dependent. Describe three mechanisms that do introduce density dependence, however trivial they may be compared with the total death rate, into human mortality.

EXERCISE 3: If you first rewrite Eq. 7 as $\int dN/[N(1 - N/K)] = r \int dt$, the equation can be solved by integration and an exact solution obtained. The trick for doing this solution is called "integration by parts," and there are three things you need to know to apply it:

1. $1/[N(1 - N/K)] = [1/N - 1/(N - K)]$
2. $\int dN/(N - N_0) = \ln(N - N_0)$
3. $\ln(a) - \ln(b) = \ln(a/b)$

Derive the exact solution, $N(t)$, to Eq. 7 by this trick, with your answer expressed in terms of r and K. There will be one constant of integration; determine its value by using the initial condition $N(t = 0) = 1$.

EXERCISE 4: Using the values $r = 0.05$ and $K = 2$, graph $N(t)$, your solution in Exercise 3, over the time range from $t = 0$ to $t = 100$.

EXERCISE 5: This exercise gives you practice using a spread sheet to solve a discrete-time approximation to our differential equation. Solve Eq. 7 numerically, with $r = 0.01$, $K = 2$, and $N(0) = 0.5$, for unit time steps from $t = 0$ to $t = 100$. Here is how to set up a spread sheet to do it:

	A	B	C
1	t	N	dN/dt
2	0	0.5	0.01*B2 – 0.005*B2^2
3	A2 + 1	B2 + C2	copy C2
4	copy A3	copy B3	copy C3
5	etc.	etc.	etc.

Here, the variable dN/dt is equal to the change in N in a time interval of $t = 1$; in other words, $dN = (dN/dt)\,dt$, but because our time step is a unit change in t, $dt = 1$. The repeated copy command will allow you to generate N values up to large values of t.

EXERCISE 6: In this exercise, you explore the validity of the assumption we made in going from Eqs. 4 and 5 to Eq. 6. You can do this exercise by solving Eqs. 4 and 5 numerically and then comparing the resulting $N(t)$ value with the $N(t)$ value obtained from Eq. 6. A spread-sheet technique similar to the one you used in Exercise 5 can be used here, but you have to modify it so that you can solve two simultaneous equations. (If you have trouble doing so, wait until you have read Problem IV-4, which teaches the method explicitly.) In solving Eqs. 4 and 5, use the numerical values $B = 0.05$, $c_x = 0.05$, $c_i = 0.1$, and $c_o = 0.2$, because they are equivalent to the r and B values used in Exercise 4. Using those values for the four parameters, solve the coupled Eqs. 4 and 5 numerically, and then make a graph of $N(t)$ obtained from the coupled equations versus $N(t)$ obtained in Exercise 4. How would you characterize the validity of the approximation? Try to find any combination of B and the c-parameters such that $r = 0.05$ and

$K = 2$ but for which the approximation would not be good. Provide a more general characterization of the conditions under which the approximation should work well.

EXERCISE 7: We derived the logistic equation by assuming that self-poisoning under crowded conditions was the factor that limited a population's growth. But that is only one of many ways in which the equation might arise. Another involves competition for a scarce resource. Let's modify our bottle by assuming that the waste products are rapidly flushed out so they do not pose a constraint, but that the food supply is not replenished. Starting with the Malthus equation, $dN/dt = rN$, let r be proportional to the amount of some nutrient in the food, such as nitrogen. Denoting the nitrogen content of the food by M, $r = r_0 M$. Rather than letting N represent the number of individuals in the population, now let it equal the amount of nitrogen in the living population. If no nitrogen is added to, or subtracted from, the bottle, then the total amount of nitrogen in food plus living bacteria must sum to a constant. Using this conceptual picture of competition for a resource, derive the logistic equation for $N(t)$.

EXERCISE 8: Returning to the Malthus equation, suppose you try to incorporate the effect of crowding by setting r, the per-capita growth-rate constant, not equal to a constant, but rather letting it be a decreasing function of time. You could, in contrast, represent medical advances by letting r be an increasing function of time. Explore the solution $X(t)$ to the Malthus equation for various assumptions about the time dependence of r. What dependence of r on t will generate a Malthus solution that is the same as the exact solution to the logistic equation you obtained in Exercise 3?

IV-3. Climate: Influence of Solar Variability

Imagine that fictional planet Earth is entirely covered by a 100-m deep, rapidly and well-mixed ocean, has no atmosphere, but in other respects (distance from sun, albedo) is just like the real Earth. What is the planet's annually averaged surface temperature? Construct a differential equation that governs the temporal behavior of the planet's surface temperature, and use it to determine how that temperature would change over time as a result of an 11-year sunspot cycle, if the sun's output changes by 1% over the course of the cycle.

. . . .

The first part of this problem repeats material from *COW-1*. There, you may recall, we wrote an energy-balance model for Earth. It read

$$\frac{\Omega}{4} = a\frac{\Omega}{4} + \sigma T^4. \tag{1}$$

In this equation, the left side is the energy flux incident upon the Earth, and the right side is the outgoing energy flux from Earth. Ω is the solar flux incident upon a surface oriented perpendicular to the sun's rays and located just above the Earth. Its numerical value is 1,370 watts/meter^2. $\Omega/4$ is also the solar flux at the Earth, but with the factor of 1/4, it is an average of the solar flux over the entire surface area of the Earth, not over a surface perpendicular to the sun's rays. Thus, the factor of 1/4 is the ratio of the area of a disk of radius R to the area of a sphere with radius R: $\pi R^2/4\pi R^2$. The term $a(\Omega/4)$ on the right side of the equation is the portion of the solar flux that is reflected off the planet. The parameter a, called the albedo, is the fraction of incident sunlight that is reflected; for the real Earth a is about 0.30.

What is the σT^4 term? Most of the energy from the sun is in the visible portion of the electromagnetic spectrum, but when it is absorbed by an object, like the Earth, this visible light is converted into heat, or as the physics books often say, infrared radiation or long-wave radiation. The rate at which this heat radiates away from the surface is governed by the Stefan–Boltzmann law, which tells us that the flux of heat is proportional to the fourth power of the surface temperature of the object. Thus, the term σT^4 is the rate at which the planet sheds the heat energy produced when the planet absorbed the fraction $(1 - a)$ of the incident solar energy. The constant σ is a very famous constant in science; its numerical value is 5.67×10^{-8} watts per (meter)^2 per (kelvin)^4. In applying this formula, you must express temperature, T, on the kelvin scale, where the freezing point of water is 273 kelvin (K) and the boiling point of water is 373 K.

The temperature T is the fourth root of the average value of T^4—that is, of T^4 averaged over the surface of the planet. Note that this term is not the same thing as the average value of T. For our watery, airless planet, T is the temperature of the surface of the ocean, because the outgoing infrared radiation would originate at that surface. On a planet with a thick-enough atmosphere, T would be the temperature of some layer in the atmosphere where the average outgoing radiation originates.

Solving Eq. 1 for T, we obtain

$$T = \left[\frac{(1-a)\Omega}{4/\sigma}\right]^{1/4} = \left[\frac{(1-0.3)(1370/4)}{5.67 \times 10^{-8}}\right]^{1/4} = 255\ \text{K}. \tag{2}$$

On the Celsius scale, this value is $255 - 273 = -18°\text{C}$, which is indeed how cold our planet would be if we lacked an atmosphere![20]

We did not need a differential equation to solve the first part of the problem because we were looking for an estimate of an average, steady-state temperature. Differential equations are used to tell us about changing conditions, and nothing was changing. For the second part of the problem, however, we need to write an equation governing how T will change if the sun's output, Ω, changes. Eq. 1 describes an energy balance, in which energy flux in equals energy flux out. If the solar output suddenly dropped, then the incoming flux would suddenly drop; the outgoing flux would not drop immediately to equal the incoming flux, however, because the outgoing infrared radiation is governed by the surface temperature, which may change only slowly. During the period when the fluxes are out of balance, there must then be a net change in the amount of energy stored on Earth because the principle of energy conservation always holds. In mathematical terms, we can express these ideas by the following equation:

$$\frac{dQ}{dt} = \frac{\Omega}{4} - a\frac{\Omega}{4} - \sigma T^4, \tag{3}$$

where Q is the energy per unit area stored in the planet. The right side expresses the imbalance of external fluxes, and the left side expresses the compensating change in the heat stored in the planet.

Eq. 3 is not very helpful, however, because the variables Q and T are different. Somehow, we have to relate them so we have a single equa-

20. This is true only if, as we assumed, the albedo is equal to its current value, 0.30. In reality, if there were no atmosphere, the albedo would be lower than 0.30; if the surface were below the freezing point and the oceans were frozen, it would be higher than 0.30.

tion for a single variable. This point is where the information about the ocean comes in. Let's assume that nearly all of the change in stored heat is the result of a temperature shift in the ocean. Because it is rapidly and well mixed, the ocean should be at a uniform temperature T. The heat content of the ocean is determined by that temperature, and the conversion factor is the heat capacity of the water.

In general, the heat energy, Q, in an object at temperature T is given by $Q = KT$, where K is the heat capacity of the object expressed in units of joules per degree. Hence, we can express dQ/dt as $K dT/dt$ (assuming K does not vary over time) and rewrite Eq. 3 as

$$\frac{K dT}{dt} = \frac{\Omega}{4} - a\frac{\Omega}{4} - \sigma T^4. \tag{4}$$

Eq. 4 is a very general, albeit simple, climate model. In many respects, it plays a role in climate modeling similar to that played by the logistic equation in population modeling. It cannot possibly capture all the complexity of a real climate system, but it is a good starting point from which to get a preliminary understanding of climate phenomena, and it is a core model to which one can add further complexity.

The constant K for an object is the specific heat of the object, in units of joules per degree per gram, times the mass of the object in grams. In our case, however, we need the heat capacity per unit area, to keep the units in Eq. 3 consistent. Thus, we need to know the heat capacity of a square meter of the ocean. The heat capacity of a column of water that is 1 m^2 in cross-sectional area and 100 m long is the mass of that column times the specific heat of water. The column of water has a volume of 10^2 m^3, or 10^8 cm^3, and thus has a mass of 10^8 grams. The specific heat of water is 4.2 joules/gm^2 °C, so the heat capacity per square meter of the ocean is given by $K = 4.2 \times 10^8$ joules/m^2 °C. Because a change in temperature of 1 kelvin equals a change of 1°C, we can also write this as 4.2×10^8 joules per $(\text{meter})^2$ per kelvin (Jm^{-2} K^{-1}). Because the units by which we have expressed Ω and σ involve watts, not joules, it will simplify things later if we express K in watt-years rather than in joules. One watt equals 1 joule/second or 3×10^7 joules/year, so $K = 14$ watt-years per $(\text{meter})^2$ per kelvin.

To solve the actual problem, let's simplify the notation somewhat by redefining new variables that will render Eq. 4 less cumbersome. We define:

$$p = \frac{\Omega/4 - a(\Omega/4)}{K} \tag{5}$$

and

$$w = \frac{\sigma}{K}, \tag{6}$$

which allow us to rewrite Eq. 4 as

$$\frac{dT}{dt} = p - wT^4. \tag{7}$$

At time $t = 0$, $p = wT^4$, or $T(0) = (p/w)^{1/4}$. This expression is equivalent to our Eq. 2 above. Numerically, $p = 17.2$ (kelvin)(year)$^{-1}$ and $w = 0.405 \times 10^{-8}$ (kelvin)$^{-3}$(year)$^{-1}$.

Now we want to see how $T(t)$ behaves if Ω, and therefore p, oscillates on an 11-year cycle with a variation in amplitude of 1%. We could, of course, solve this question numerically on a spreadsheet, but it will be very useful and insightful to obtain an approximate analytical answer. To do this, we will take advantage of the fact that T is expressed on the kelvin temperature scale. On this scale, everyday temperatures, and even the subfreezing value we obtained for T (Eq. 2), involve numbers that are in the rough range of 250 to 300 kelvin. Intuitively, we would be surprised if the sunspot cycle induced as much change in T as, say, the difference in temperature between summer and winter or day and night. Those temperature swings are on the order of magnitude of 10°C or 10 kelvin, which is small compared with the ambient average value we obtained of 255 kelvin.

Because of this comparison, we can simplify Eq. 7 with a fine approximation. We set the actual temperature we wish to solve for, $T(t)$, equal to its average value (255) plus a correction, which we will call $X(t)$:

$$T(t) = 255 + X(t), \tag{8}$$

where T and X are in kelvin, and thus $X \ll 255$. The quantity T^4 in Eq. 7 can now be written, using a power series expansion, as $(255 + X)^4 = (255)^4 + 4(255)^3X +$ terms involving $(255)^2X^2$, $255X^3$, and X^4. These additional terms on the right side are all relatively much smaller than the first two, and thus we can ignore them as a reasonable approximation.

Eq. 7 then becomes

$$\frac{dX(t)}{dt} = p - w[(255)^4 + 4(255)^3X(t)]. \tag{9}$$

Now we have to incorporate in our equation the information about the sunspot cycle. We do this by writing $p(t)$ in terms of its mean value, p_0, which numerically is 17.2 (kelvin)(year)$^{-1}$, and its oscillation around that mean. The oscillation has a period of 11 years and an amplitude of 1% of p_0, so with t in units of years, we can write p in the form

$$p(t) = p_0\left[1 + (0.005)\sin\frac{2\pi t}{11}\right]. \tag{10}$$

You should convince yourself that this function has the desired properties of an 11-year periodicity, an amplitude of 1% of p_0, and a mean of p_0.

Substituting Eq. 10 into Eq. 9, we get

$$\frac{dX(t)}{dt} = p_0 + 0.005p_0 \sin\frac{2\pi t}{11} - w(255)^4 - w(255)^3 X(t). \tag{11}$$

But $p_0 = w(255)^4$, so this equation simplifies to

$$\frac{dX(t)}{dt} = 0.005p_0 \sin\frac{2\pi t}{11} - w(255)^3 X(t). \tag{12}$$

Substituting numerical values for p_0 and w, we can re-express the equation as

$$\frac{dX(t)}{dt} = 0.086 \sin\frac{2\pi t}{11} - 0.067X(t). \tag{13}$$

This equation can be solved exactly. It is of the general form

$$\frac{dX}{dt} = \beta(t) + \alpha(t)X(t), \tag{14}$$

where β and α are arbitrary functions of t. Eq. 14 has an exact solution given by Eq. 14 of Problem IV-1. Applying that solution to our Eq. 13 above, we get

$$X(t) = ce^{-0.067t} + 0.017 \sin\frac{2\pi t}{11} - 0.145 \cos\frac{2\pi t}{11}, \tag{15}$$

where c is the constant of integration.

At large time t, the first term on the right side of Eq. 15 is very small; because that is the only term containing the constant of integration, our solution approaches a function independent of the intitial condition. So let's arbitrarily assume that $X(0) = 0$, thereby giving us $c = 0.145$. Figure IV-1 shows the behavior of $X(t)$. After about 50 years, the transient behavior of $X(t)$, which was a consequence of our initial condition, has damped out, and it has settled into a simple periodic function with a period of 11 years. The amplitude of the cycle is 0.3°C. As seen in the figure, and as calculable from Eq. 15 (Exercise 2), the minimum value of X occurs at $t \sim 11n$, where n is any integer. We note from Eq. 10 that at these values of t, the solar cycle is not at a minimum but rather at its average value of p_0. Thus, the temperature fluctuation is out of phase with the sunspot cycle, lagging by a quarter of a cycle.

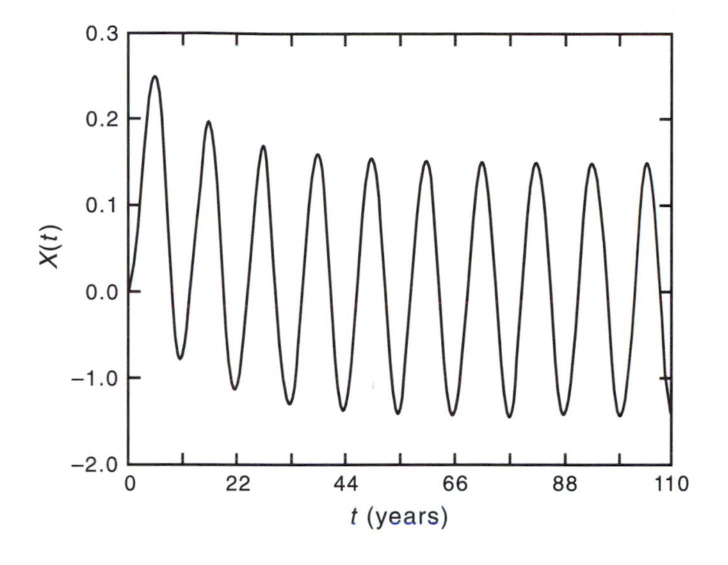

Figure IV-1 The deviation of globally and annually averaged surface temperature from its long-term average (255 kelvin) under an 11-year solar sunspot cycle with an amplitude of 1% of average solar output.

EXERCISE 1: Classify Eq. 7, with p equal to a constant, and Eq. 12 according to our taxonomic key to the differential equations.

EXERCISE 2: Calculate the values of t for which X as given by Eq. 15 is minimum, assuming that t is very large so that the term $ce^{-0.067t}$ can be neglected. Explain in words why the temperature cycle is out of phase with the sunspot cycle.

EXERCISE 3: Recalling what we said in the paragraph preceding Eq. 2, T is not the average surface temperature of the planet but rather the fourth root of the average of T^4. What is the difference? Is the average value of T going to be bigger or smaller than 255 kelvin? Can you estimate the average value?

EXERCISE 4: Over the past 100 years, the Earth's average surface temperature has increased by about 0.6 kelvin. Some have suggested that this warming trend is the result of an upward trend in solar output. Although solar output was not measured highly accurately 100 years ago, an increase in solar output of more than a few tenths of a percent is unlikely to have occurred during the past century. Neglecting the sunspot cycle and examining just the effect of a steady upward trend in solar output, determine whether an increase in average solar output of, say, 0.5%/century accounts for a 0.6 kelvin/century warming.

EXERCISE 5: If p is a constant, Eq. 7 can actually be solved exactly. The trick is the same one you used in Exercise 3 of Problem IV-2 to solve the logistic equation—integration by parts. Paralleling what you did there, you first have to factor the expression $(T^4 - p/w)$ into a product of four terms of the form $(T - a_1)(T - a_2)(T - a_3)(T - a_4)$, where the a_i are the four fourth-roots of p/w. Then you will have to re-express $1/(T^4 - p/w)$ as a sum of $b_i/(T - a_i)$, where the b_i are suitable constants. Finally, you can integrate each of these terms. The only new ingredient here is that two of the fourth-roots of p/w are imaginary numbers. You will want to make use of the identity: $\arctan(x) = (i/2)\ln[(i + x)/(i - x)]$ where $i = \sqrt{-1}$. Find the exact solution in terms of an arbitrary initial condition $T(0)$.

IV-4. Community Ecology: Origins of Cycles

A population of foxes and rabbits coexist in an ecosystem and are observed to undergo more or less periodic booms and crashes, with fluctuating amplitude. The period is not 1 year, nor is it the average El Niño recurrence interval or any other value that might have an obvious external cause. Explain the origin of the cycles.

. . . .

The situation described here is the classic predator–prey configuration that community ecologists enjoy modeling. Often, they add more populations to the model, for example, a population of grass for the rabbits to feed upon, several species of competing grass-eaters, several species of competing plants, two species of predators that eat rabbits, or combinations of those added complexities. By studying the simple case of rabbits and foxes, however, many of the important concepts will emerge, and you will be on your way toward tackling more complicated modeling exercises.

To begin, we must choose appropriate variables. We will take $F(t)$ to be the total biomass of the living foxes. Similarly, we will take $R(t)$ to be the total live rabbit biomass.[21]

The rate of change of a variable Y is the difference between the rate of input to Y, or Y_{in}, and the rate of output, Y_{out}:

$$\frac{dY}{dt} = Y_{\text{in}} - Y_{\text{out}}. \tag{1}$$

The next step is to determine the inputs and outputs to R and F, the stocks of rabbit and fox biomass. We will assume F increases solely as the result of foxes eating rabbits and decreases because foxes die of disease and "old age." To begin, we will take the loss of fox biomass to be linearly proportional to F, but later we will see that we have to modify that assumption. We will start by assuming that rabbit biomass decreases only because of predation by foxes (no disease or other causes of death) and that the net difference between growth and nonpredatory death of rabbits occurs at a constant per-unit-biomass rate

21. We could take F and R to be the number of foxes and rabbits, but it is generally easier to work with biomasses. There are two reasons. First, if we were including grass in the model, we would want to use grass biomass; counting grass blades is no fun, and thus it would be easier to use biomass if we want to compare model predictions to field data. Second, when foxes eat rabbits, the relationship between the gain in fox biomass and the loss in rabbit biomass is easier to conceptualize (see below) than is the relationship between the gains and losses in numbers of individuals.

of growth. Thus, in the absence of foxes, rabbit biomass would obey the Malthus equation; that assumption, too, will be modified later.

To describe the relation between loss of rabbit biomass and gain in fox biomass, we turn to general insights about animal physiology. When a rabbit is caught by a fox, all of that rabbit's biomass is lost from R. But the gain in F is not equal to the loss of R, even when the entire rabbit is eaten, because of excretion and metabolism losses. Typically, animals can convert only a fraction, e, of their food into new biomass. The fraction e is an efficiency factor; measured e values for various predator–prey pairs of species range from less than 0.01 to as high as 0.3 for some zooplankton feeding on phytoplankton. We will take $e = 0.02$ for our foxes and rabbits.

The next step in constructing a model is to determine mathematical equations that correspond to these verbal descriptions of the inputs and outputs. By our assumptions, the net growth of rabbits, in the absence of predation, is proportional to the rabbit biomass, R, with the proportionality factor r representing the per-unbit-biomass net rate of rabbit biomass increase:

$$R_{\text{in}} = rR(t). \tag{2}$$

In keeping with the spirit of this book, we want to represent the interaction between the rabbits and the foxes as simply as possible, compatible with our knowledge. If the rabbit or the fox population vanished, then no rabbits would be eaten. Thus, the interaction term must vanish if either R or F vanishes. The simplest mathematical expression that conveys that information is a product, RF. Thus, we take

$$R_{\text{out, predation}} = hFR, \tag{3}$$

where h is a predation constant characterizing the strength of the interaction.

The inflow of biomass to the fox population will be similar to Eq. 3 but reduced by the efficiency factor, e:

$$F_{\text{in, predation}} = ehFR. \tag{4}$$

And finally, we describe the loss of fox biomass by sickness and other causes by the density-independent expression

$$F_{\text{out, death}} = qF. \tag{5}$$

Putting all this together, we arrive at the following pair of coupled differential equations:

$$\frac{dR}{dt} = rR - hFR \tag{6}$$

"I think I just found the monkey wrench in our fox model. . ."

and

$$\frac{dF}{dt} = ehFR - qF. \tag{7}$$

In Chapter 5, you will learn some analytical techniques for studying properties of coupled nonlinear equations without obtaining numerical solutions; for now, it will be most insightful to examine numerical solutions.

To solve Eqs. 6 and 7 numerically, we have to select numerical values for the parameters r, q, h, and e. We want only reasonable values here, because we are not fitting real data; in the exercises, you will get a chance to explore the sensitivity of our results to the numerical value of the parameters. On the reasonable assumption that rabbit biomass has a turnover time of roughly 3 or 4 years, we take $r = 0.3/\text{y}$. Similarly, a fox biomass turnover time of about 5 years leads to $q = 0.2/\text{y}$. Without loss of generality, we can take $h = 1$ because that just sets the unit with which we measure the biomass of rabbits and foxes.

To determine a value for e, we assume for the moment that R and F are in steady state: $dR/dt = dF/dt = 0$. In a putative rabbit steady state, Eq. 6 then implies that $F = r/h = 0.3$. If $dF/dt = 0$, then $R = q/eh$. The value we choose for e, the efficiency of converting rabbit biomass into fox biomass, then determines the ratio F/R. In particular, $F/R = (r/h)/(q/eh) = er/q = 0.3e/0.2 = 1.5e$. We know that a population of foxes cannot sustain itself on a comparably sized population of rabbits; we must have $R \gg F$. The choice $e = 0.02$ is a reasonable efficiency factor for small mammals and gives us a ratio of $F/R = 0.03$.

With these choices, our equations now look like

$$\frac{dR}{dt} = 0.3R - RF \tag{8}$$

and

$$\frac{dF}{dt} = 0.02RF - 0.2F. \tag{9}$$

All that stands between us and a numerical solution is a choice of initial conditions and a procedure for solving the equations. We arbitrarily pick the values $R(0) = 9.0$ and $F(0) = 0.25$, but Exercise 3 gives you a chance to explore the sensitivity of our findings to the initial conditions. To solve Eqs. 8 and 9, we will use a spread-sheet approach as in Exercise 5 of Problem IV-2.

The table below illustrates how the spread sheet is set up to solve the coupled differential equations. We created a column of time steps by using the copy command to increase t in each row by a factor of 0.01. The choice of 0.01 was made after a little trial and error; in Exercise 2, you will explore the consequences of making the time step larger (which saves time but runs the risk of generating wildly fluctuating output that does not actually resemble the solution to the differential equations) or smaller (which eats up a lot of memory and can overload a typical personal computer).

Rows 1 and 2 are entered by hand. On row 3, after columns A, B, and C are entered by hand, the copy command is used to generate D3 and E3. Then, starting on row 4, the entry for all the columns is generated by copying the row above, all the way down to the largest t value for which output is desired.

	A	**B**	**C**	**D**	**E**
1	*t*	*R*	*F*	*dR/dt*	*dF/dt*
2	0	9	0.25	0.3*B2 – B2*C2	0.02*B2*C2 – 0.2*C2
3	A2 + 0.01	B2 + D2*0.01	C2 + E2*0.01	copy D2	copy E2
4	copy A3	copy B3	copy C3	copy D3	copy E3
5	etc.	etc.	etc.	etc.	etc.

If you set up the spread sheet correctly, the first few rows should look like this:

T	*R*	*F*	*dR/dt*	*dF/dt*
0	9	0.25	0.45	–0.005
0.01	9.0045	0.24995	0.450675	–0.00498
0.02	9.009007	0.2499	0.451349	–0.00495
0.03	9.01352	0.249851	0.452022	–0.00493

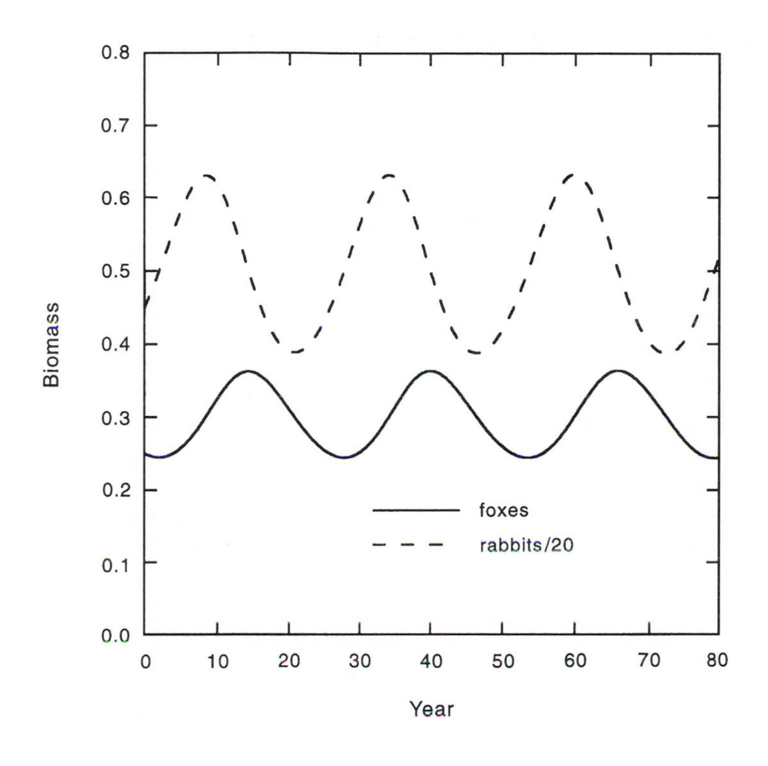

Figure IV-2 Population cycles for rabbits and foxes in a simple, deterministic Lotka–Volterra model without density dependence.

Figure IV-2 shows a graph of both $R/20$ and F versus time, for $t = 0$ to $t = 80$ years (we divide R by 20 for purposes of showing both species on the same graph). It illustrates the classic pattern of the predator–prey cycle:

- An increase in the number of rabbits permits an increase in the number of foxes.
- As more foxes eat more rabbits, the number of rabbits declines.
- That, in turn, causes the number of foxes to decline.
- That allows the number of rabbits to increase.

Note that the cycle is remarkably smooth and persistent, that the period of the cycle is about 25 years, and that the fox biomass peaks about a quarter period after the rabbit biomass peaks.

Eqs. 6 and 7, together, are called the Lotka–Volterra equations for a predator and a prey. The model they describe is arguably the "core model" in community ecology, in the sense that a large proportion of more complicated (and more importantly, more realistic) models use these equations as the starting point, adding on more dynamic mechanisms and more species as appropriate.

Several features of this core model are relatively independent of the choice of parameters r, q, h, and e. In particular, if the initial conditions for the biomasses of the two populations are chosen to be fairly near the steady-state solution ($R = q/eh$, $F = r/h$), then the quarter-period difference between the peak values for R and F is a robust characteristic of the equations. The period itself does depend on the choices of the parameters, however, and the amplitude of the cycle depends on the choice of initial conditions $R(0)$ and $F(0)$. In the exercises, you will get a chance to explore some of these dependencies.

With suitably chosen parameters r, q, h, and e, the cycles described by this simple core model often resemble rather well the shape of actual population cycles observed in nature. Even the quarter-period phase difference is often observed in real cycles. So why are further complexities necessary? The reason is that despite appearances, the core model is pathological. A hint of that pathology is the observation that the amplitudes of the cycles depend on initial conditions. Real populations are always buffeted by forces that, in effect, keep changing the initial conditions; the Lotka–Volterra cycles would, under these circumstances, continually be resetting the amplitude and phase. Much like a ball rolling on a flat table, the Lotka–Volterra orbits show no persistence under perturbations. This reflects an underlying lack of stability of the model. The mathematical tools needed to evaluate the stability of models like Eqs. 6 and 7 are presented in the next chapter, but even without the analytical tools, you can see from numerical analysis that the solutions to the equations for a variety of initial conditions do not converge toward a stable state (Exercise 3).

There is a simple, realistic way to add a term to Eqs. 6 and 7 and thereby stabilize the solutions, that is, to make them insensitive to initial conditions. Moreover, this "fix" corresponds to something observed in nature, namely, the type of density-dependent death mechanism discussed in Problem IV-2, where we presented the logistic model. Let's explore the consequence of adding a density-dependent term to the "death-of-foxes" term in Eq. 9. To do so, we replace the linear death rate, qF, with a term that has linear plus quadratic dependence on F:

$$qF \rightarrow q_1F + q_2F^2. \tag{10}$$

A similar expression could be added to the rabbit equation. Figure IV-3 shows the effect of adding such a density-dependent death rate to our core model, keeping the initial conditions and all other parameters the same as in our previous case, but replacing

$$0.2F \rightarrow 0.05F + 0.5F^2 \tag{11}$$

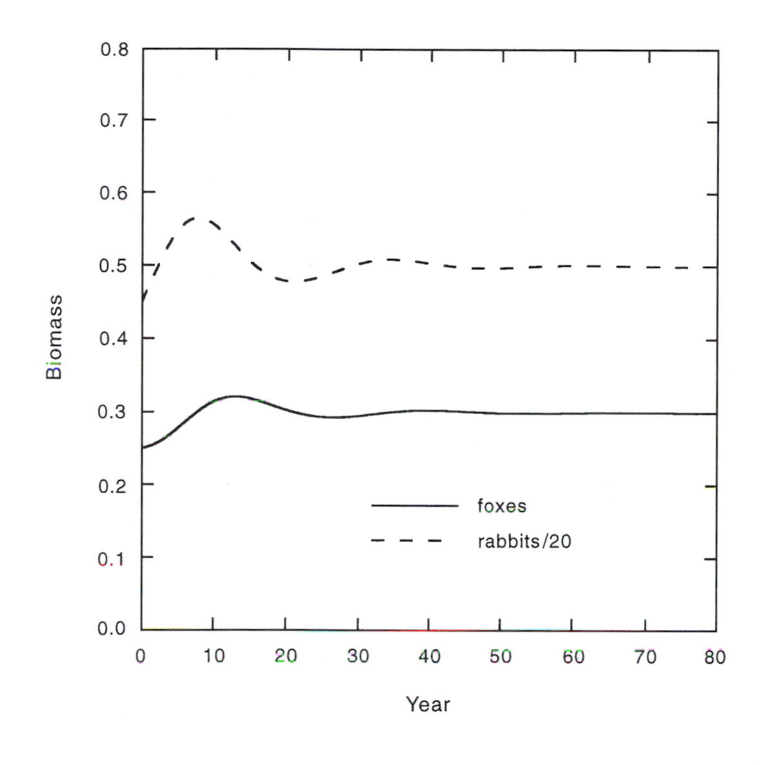

Figure IV-3 Biomasses of rabbits and foxes in a Lotka–Volterra model with density dependence; note that the cycles in Figure IV-2 have damped out.

with q_1 and q_2 chosen here to preserve the value of the death rate at the steady-state solution $F = 0.3$. The full equations now read

$$\frac{dR}{dt} = 0.3R - RF \tag{12}$$

and

$$\frac{dF}{dt} = 0.02RF - 0.05F - 0.5F^2. \tag{13}$$

As illustrated in Figure IV-3, the solutions to Eqs. 12 and 13 no longer exhibit persistent cycles; rather, the trajectories damp out and approach the steady-state values $R = 10.0$ and $F = 0.3$. You can easily show (Exercise 4) that for any set of initial conditions, this same behavior of a damped approach to $R = 10$ and $F = 0.3$ is obtained. A similar behavior would occur if we insert the density dependence into the dR/dt equation instead of, or in addition to, the dF/dt equation.

Thus, by augmenting the Lotka–Volterra equations with density-dependent death rates, we have solved one problem (stability against variation in initial conditions), but we are now left with another problem. In particular, we have lost those cycles that were an appealing feature of Eqs. 8 and 9.

Can we have our cake and eat it, too? That is, can we retain the stability and greater mechanistic realism of the equations that arise with the introduction of density-dependent death rate and yet preserve the nice, cyclic features of the solutions that arose in the absence of density dependence? We can. One way to accomplish this is to add one more feature to the model that is also "out there" in nature: stochasticity, or random variability.[22]

Stochasticity can take many forms, from differences in rate constants among individuals in a population that arise by chance, to random variation over time in the average rate constant for the entire population. Eqs. 14 and 15 incorporate into our model the latter type of stochasticity, which can be thought of as a consequence of interannual climate variability. These equations are identical to Eqs. 12 and 13 except that random-number-generating functions have been inserted in the terms describing the linear net growth rate of rabbits and the linear death rate of foxes:

$$\frac{dR}{dt} = 0.6\,(\mathrm{RAND1})\,R - RF \tag{14}$$

and

$$\frac{dF}{dt} = 0.02RF - 0.1\,(\mathrm{RAND2})\,F - 0.5F^2. \tag{15}$$

RAND1 and RAND2 are functions that, independently of each other, generate random numbers between 0 and 1 at every time step. Because the time-average of these functions is 0.5, we have replaced 0.3 in Eq. 12 by 0.6 in Eq. 14, and 0.05 in Eq. 13 with 0.1 in Eq. 15. Most spreadsheet programs, such as EXCEL, provide a simple means of inserting random-number generators into calculations.

Figure IV-4 illustrates the effect of this stochasticity on rabbit and fox biomass. In the absence of the random-number generators, the solutions would, as we saw, damp out to steady state. The figure shows that stochasticity restores the cycles of the original density-independent core model, although they are no longer as smooth. The solutions possess the nice cyclic properties of Eqs. 8 and 9, including even the quarter-period phase difference, but the density dependence of Eq. 15 retains the greater mechanistic realism of Eqs. 12 and 13 and

22. Nesbet, R., Gurney, W. 1977. A simple mechanism for population cycles. *Nature* 263:319–21.

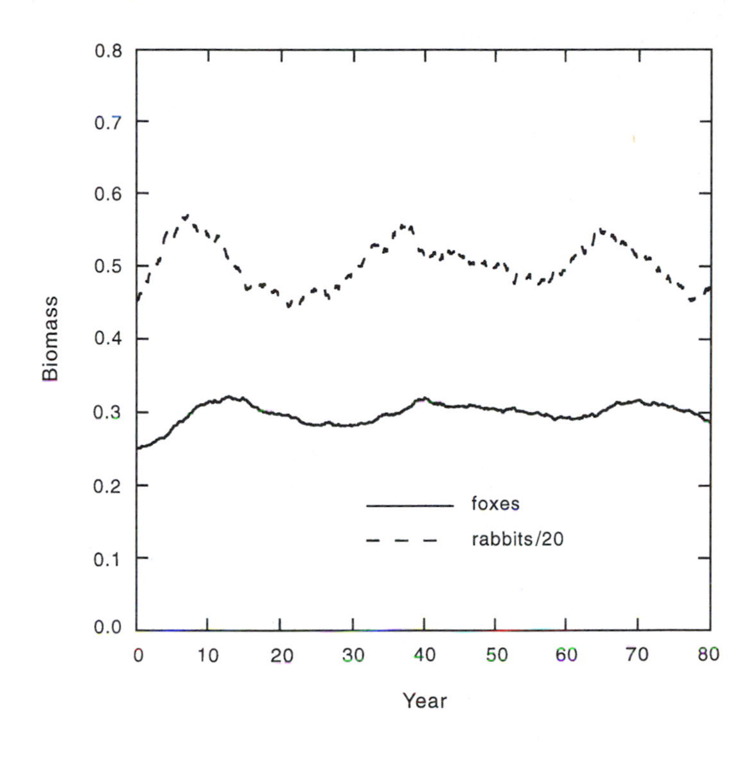

Figure IV-4 Effect of stochasticity in a density-dependent Lotka–Volterra model of rabbits and foxes.

provides at least some of the insensitivity to initial conditions of Eqs. 12 and 13.

The Lotka–Volterra equations can also be generalized by adding more species to the system. For instance, if we let each species be labeled by an index i, so that X_i = the biomass of the ith species, then

$$\frac{dX_i}{dt} = \sum_j \beta_{ij} X_i X_j + \alpha_i X_i + \gamma_i X_i^2. \tag{16}$$

If a pair of species satisfies the condition (sign $\beta_{ij} = -$sign of β_{ji}), then that pair is a predator–prey pair, whereas if the pair satisfies the condition (sign β_{ij} = sign β_{ji}), then the pair is competitive if the sign is negative or mutualistic if the sign is positive. Eqs. 16 are sometimes referred to as the generalized Lotka–Volterra equations.

Even these equations can be further complicated and perhaps made more applicable to nature. One such change involves the interaction term $\beta_{ij} X_i X_j$, which can be altered to take into account the phenomenon of food saturation. For example, as the rabbit population increases, foxes eventually are no longer limited by the number of rabbits. To incorporate this interaction into a model, we need to replace the function βRF with a new interaction function that has the property that as R increases, the function becomes increasingly independent of R; in other words, in the limit $R \to \infty$, the interaction function approaches βFK. At small R values, the interaction function should look like our original expression βRF. A simple and widely used example of such an interaction function is described by the replacement,

$$\beta RF \to (\beta RF)\frac{K}{K+R}. \tag{17}$$

We might also make some of the terms in Eqs. 16 be either random or dependent upon the solar cycle. For example, the dependence of photosynthesis on sunlight might be expressed by replacing an appropriate αX term (where X corresponds to a species of plant) by $\alpha[1 + \lambda \sin(2\pi t)]X$. Here, t is measured in years, $t = 0$ corresponds to the spring equinox, and the ratio of the winter-solstice photosynthesis rate to the rate at the equinox is $(1 - \lambda)$. Exercise 5 gives you a chance to see whether a one-year solar cycle can excite from the damped (density-dependent) Lotka–Volterra model a cyclic behavior of period different from one year.

Amidst the somewhat bewildering variety of ways to expand upon the simple core model in Eqs. 6 and 7, you might well wonder what general conclusions emerge from all of this. Remarkably, some useful generalities emerge from the study of numerous special cases of Eqs. 16 that can even be proved for the general case. If there is no density dependence, cycles result, but they always possess the unreal features discussed above. If density-dependent death rates are introduced, the solutions damp out to steady states. Those steady states are stable[23] if all the interactions are of the predator–prey type, but they can become unstable if sufficiently strong competitive or mutualistic interactions are present. When stochastic or other other explicit exogenously time-dependent parameters are inserted along with density-dependent death rates, the cycles are excited, and some of the attractive features of the core model are regained, along with some of the insensitivity to initial conditions of the density-dependent models.

There are other ways to achieve cycles without introducing explicit time-dependence as we did here. For example, models that incorporate predator and prey saturation mechanisms yet have constant parameters can yield "limit cycles"; these cycles are insensitive to initial conditions and persist over time. Some cycles observed in nature have much larger amplitudes than those emerging from our simple rabbit–

23. See Chapter 5 for the technical definition of stability; for our purposes here, stability means insensitivity to initial conditions.

fox models, and these may result from adding to our core model more complex, multispecies, nonlinear interactions. A more complete analysis of ecological cycles is beyond the scope of this book, but I urge interested readers to explore further by turning to the books listed in Further Reading. In the meantime, the exercises below give you a chance to explore on your own some of the issues raised here and to discover, numerically, general truths about various formulations and ramifications of the core model.

EXERCISE 1: Classify each of the pairs: Eqs. 8 and 9; Eqs. 12 and 13; and Eqs. 14 and 15 within our taxonomy.

EXERCISE 2: Get the model expressed by Eqs. 8 and 9 up and running on a spread sheet, and verify the results shown in Figure IV-2. Explore the sensitivity of the numerical solutions to the size of the time step by examining output for time steps of 0.1 and 0.001. What can you conclude?

EXERCISE 3: Again using Eqs. 8 and 9, explore sensitivities of the amplitudes and relative phases of the cycles to parameters $r, q, h, e, R(0)$, and $F(0)$. Try to use dimensional consistency, as in Problem III-3, to determine the parameter combinations that control the period and amplitude of the cycles. [*Hint*: Recall that the variables R and F have units of biomass, as do both the ampitude of the cycles and the differences between initial and steady-state values of R and F.]

EXERCISE 4: Using the modification of Eqs. 8 and 9 described by Eqs. 12 and 13, explore the effect on the time dependence of R and F of various values for the ratio of q_1/q_2. In each case, keep the steady-state value of F the same, so that $q_1F + q_2F^2 = 0.06$. How do your results depend upon initial conditions? Does the value of q_1/q_2 cause a qualitative change in the output? Try to draw a generalization about the dependence of model output on this ratio.

EXERCISE 5: Construct and explore the behavior of a model for a plant and an herbivore, in which the growth of the plant is governed by a solar cycle. If the death rate of either the plant or the herbivore is density-dependent, can the solar cycle restore the cyclic behavior of the undamped (density-independent) model?

EXERCISE 6: Examine the behavior of a Lotka–Volterra model (Eqs. 16) in which two competing species of herbivore each graze on grass (a three-variable model). Choose your own combinations of parameters, and explore possible behaviors of the system.

EXERCISE 7: Examine the behavior of the deterministic rabbit–fox model (Eqs. 8, 9) with the term RF replaced by $RFK/(K + R)$ as in Eq. 17. Choose a few values of the parameter K, and explore possible behaviors of the system.

IV-5. Biogeochemistry: Nitrogen Fertilization and the Carbon Sink

> When nitrogen in the form of anthropogenic nitrate or ammonium is deposited on soils, both plant growth and microbial activity can be stimulated. That growth stimulation, in turn, can alter the net flow of carbon between the atmosphere and the ecosystem, leading to a gain or loss of stored carbon. Construct a model that can be used to explore the effect of anthropogenically deposited nitrogen on carbon storage in a terrestrial ecosystem, and use it to characterize the conditions under which such nitrogen input could result in a large net flow of carbon from the atmosphere to plants or soils in terrestrial ecosystems.

• • • • • • •

Biogeochemistry is the study of the stocks and flows of elements within an ecosystem or within the global biosphere. The "bio" in biogeochemistry reflects two facts: (a) many of the elements of interest are essential to life, and (b) the flows of many elements, essential or not, result from the activity of microorganisms. Biologically essential elements, such as carbon, nitrogen, and phosphorus, cycle among a variety of components of the biosphere, such as living matter, soil, water, and the atmosphere; in the process, they pass back and forth among a variety of chemical forms. Because biogeochemical phenomena involve microbial populations, interacting species of plants and animals, and chemical flows, it should not be surprising that models for describing such phenomena will involve aspects of the models explored in Problems IV-1, IV-2, and IV-4.

Here, we will develop and analyze the fifth of our core models, one that can be used to explore several issues involving stocks and flows of critical elements in the biosphere. Recall that in Problem IV-4, our initial, simplest, core model for a predator–prey system (the Lotka–Volterra equations) was instructive but had to be augmented to provide a reasonable description of actual predator–prey systems. Here, too, we will begin with the simplest possible core model and then learn how to augment it to provide more reliable insight into the nature of biogeochemical cycles.

Before diving right into differential equations, however, let's look at some actual numbers describing nitrogen and carbon and do a simple "back-of-the-envelope" estimation to see why the issue raised in the problem statement could, in principle at least, be globally important.

The global anthropogenic flux of fixed nitrogen (i.e., nitrogen in the form of nitrate or ammonia, which are the major forms that plants can use as fertilizer) to forested ecosystems is currently about 5×10^6 tons(N)/year. This flux is largely composed of ammonium (NH_4^+) and nitrate (NO_3^-) produced industrially for agriculture and of nitrate produced from the nitrogen oxides emitted during fossil-fuel burning. The anthropogenic flux is comparable in magnitude to the flux of fixed nitrogen from nonanthropogenic (natural) processes.

Suppose all of this added anthropogenic nitrogen were to be incorporated into new plant growth. Trees consist largely of woody tissue, which has a very high carbon-to-nitrogen ratio (typically $C/N \sim 300$), and foliage, which has a lower ratio ($C/N \sim 20$). If all the added nitrogen ended up in the form of new wood, then the amount of additional carbon stored each year in the form of wood as a result of the anthropogenic nitrogen flux would be $\sim 300\ (5 \times 10^6) = 1.5 \times 10^9$ tons(C)/year. This is a large amount of newly stored carbon, equal to about one-fourth of the world's annual carbon emissions from fossil-fuel burning and roughly equal to the amount of "missing carbon".[24]

Unfortunately, this estimation could be misleading for many reasons. First, much of the deposited nitrogen may volatilize to the atmosphere or wash away into the oceans, in which case much less than 1.5×10^9 tons(C)/year would be sequestered. Second, not all the added nitrogen that does remain in the forest will end up in trees; some may be transformed into soil organic matter after passing through various stages of biogeochemical cycles. Third, of the nitrogen that ends up in trees, not all may reside in woody parts. Because the carbon-to-nitrogen ratio (C/N) of soil organic matter and of the non-woody parts of trees is typically in the range of 20 to 30 rather than $\sim$300, the amount of sequestered carbon will be greatly reduced to the extent the additional nitrogen does not reside in wood.

As we have seen, a useful first step in constructing a differential equation model is to identify the key variables the model should incorporate. Both carbon and nitrogen can exist in two forms: inorganic and organic. Carbon dioxide and carbonates are examples of inorganic carbon-containing substances; salts of ammonium, nitric acid, and nitrogen gas are examples of inorganic nitrogen-containing substances. Organic forms of carbon and nitrogen are found in living organisms, as, for example, in the amino acids that are among the basic chemical

24. "Missing carbon" is the difference between the amount of carbon emitted to the atmosphere each year from human activities and the sum of two quantities: the annual increase in the atmospheric carbon content and the amount of carbon estimated to disappear each year from the atmosphere by ocean or terrestrial uptake. Missing carbon is thus a name for our ignorance about mechanisms and rates of removal of carbon dioxide from the atmosphere (see Problem IV-1). Our back-of-the-envelope calculation indicating that 1.5×10^9 tons(C) could be taken up as a result of nitrogen fertilization of forests suggests the possibility that the missing carbon resides in additional woody matter in forests.

building blocks of life, and in dead organisms that have not yet decomposed to an inorganic (or "mineral" as it is sometimes called) form. Such dead organisms include not only leaf litter on the forest floor, but the soil organic matter (for example, humus) that results from the partial decomposition of dead animals and plant matter. Our model will have to include both inorganic and organic forms because organic matter is synthesized (for example, in photosynthesis) from inorganic matter and is decomposed back to inorganic forms in the course of biogeochemical cycling.

Because biogeochemical models describe the stocks and flows of substances like nitrogen, you might be tempted to think that the form of such models would simply resemble our pollutant stock and flow model in Problem IV-1. Were this so, we would write down a linear, donor-controlled model for the time derivative of nitrogen in, say, soil. The production of inorganic matter would be proportional to the amount of organic matter in the soil; the organic nitrogen would be the chemical feedstock, and the organic carbon would be necessary to power the microbes that mediate the process. Because plants take up inorganic nitrogen when they grow, it would also be reasonable to include in the model a predator–prey type of interaction, as in Problem IV-4, in which nitrogen is the prey and plants are the predator.

Our core model will indeed contain these two features, but that is only a beginning. Biogeochemical cycles generally include a critical feature absent from either pollutant stock–flow or predator–prey dynamics: microbially mediated transformations. As a consequence of microbial mediation, the rate of production of inorganic nitrogen in soil is dependent upon the living microbial community as well as upon the amount of organic matter in soil. Moreover, the very growth of microbes is dependent upon both organic carbon as an energy source and inorganic nitrogen for protein building. As we will see, these novel features of biogeochemical processes can result in novel equations that are not simply combinations of the models in Problems IV-1 and IV-4.

Figure IV-5 shows in simplified form the dominant structures of the nitrogen and carbon cycles in a terrestrial ecosystem. The nitrogen cycle involves plants, soil, and soil organisms (mostly microoganisms) that transform organic nitrogen in soil to inorganic (fertilizer) form in which nitrogen can be used by plants. In the carbon cycle, the inorganic carbon pool is largely in the atmosphere in the form of carbon dioxide, which is the source of carbon for plants.

There are many variables we could include in our biogeochemical model, but as in all uses of models, tough choices have to be made. A full list of possible variables and processes would be effectively endless, and so we must rein in our desires if we are to get on with the task of model building. To be systematic about these choices and to lay out in advance the successive increases in complexity that will characterize the sequence of models introduced here (starting with the simplest

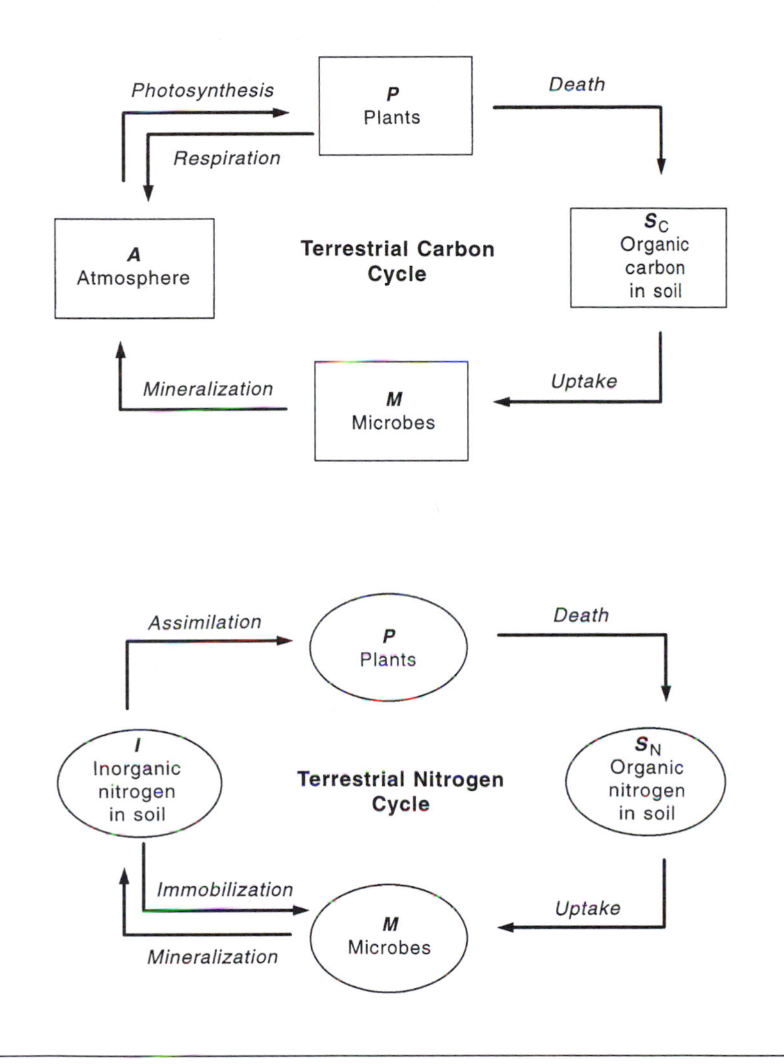

Figure IV-5 The main features of terrestrial carbon and nitrogen cycles.

core model), consider the following list of possible approximations:

1. Ignore time variation in C/N of soils
2. Ignore time variation in microbial populations
3. Ignore the differences in C/N of the separate parts of plants (leaves, trunk, fine branches, fine roots, etc.)
4. Ignore losses of nitrogen or carbon from the system by erosion, leaching, or volatilization
5. Ignore changes in the atmospheric level of carbon dioxide
6. Ignore the distinction between nitrate and ammonium in soil

7. Ignore nitrogen in its various forms in the atmosphere
8. Ignore the nitrogen and carbon contents of the different soil organic matter fractions (older and less decomposable versus younger and more decomposable organic forms of nitrogen and carbon)
9. Ignore all the different roles of distinct types of soil organisms (bacteria vs. fungi; symbiotic vs. nonsymbiotic fungi; denitrifying vs. nitrogen-fixing vs. nitrifying bacteria; microoganisms vs. larger soil organisms)
10. Ignore nitrogen fixation (conversion of molecular nitrogen gas to ammonia)
11. Ignore denitrification (conversion of nitrate to molecular nitrogen or nitrous oxide)
12. Ignore the roles of water, sunlight, temperature, and trace elements and other nutrients besides nitrogen that influence plant growth.

Our first and simplest (core) model presented below is based upon every approximation in the list above! It is a starting point, a framework on which more detail can readily be hung later. In Models 2 and 3, respectively, we will relax the first two and then the third of the listed simplifications.

Model 1. The variables included in our core model are the organic nitrogen contents of the living plants (P), the organic nitrogen content of the soil (S_N), and the inorganic nitrogen content of the soil (I). Because of approximation 1, we need not include carbon variables; because of approximation 2, we need not introduce a dynamic variable for the microorganism population; because of approximation 3, we include only a single variable for plant nitrogen.

Approximations 4, 10, and 11 imply that our model system is closed—the sum of the nitrogen contents of the three nitrogen pools, P, S_N, and I, is a fixed contant, T:

$$T = P + S_N + I. \tag{1}$$

The differential equations describing our core nitrogen cycle are as follows:

$$\frac{dP}{dt} = rPI - fP^2 \tag{2}$$

$$\frac{dI}{dt} = bS_N - rPI \tag{3}$$

$$\frac{dS_N}{dt} = fP^2 - bS_N \tag{4}$$

In these equations, the terms have the following meanings:

rPI is the rate of uptake of inorganic nitrogen by plants;
fP^2 is the loss of plant nitrogen as leaves are shed and trees die [we have assumed a pure nonlinear (or density-dependent) form for this loss rate, but in Exercise 1 you can explore the implications of including a linear dependence on P for this loss rate];
bS_N is the rate of inorganic nitrogen formation resulting from microbial processing of organic matter.

Note that by virtue of the form of these equations,

$$\frac{d(P + I + S_N)}{dt} = 0, \tag{5}$$

and thus the conservation-of-nitrogen condition expressed in Eq. 1 is automatically satisfied.

To determine what Model 1 can tell us about the effect of deposited nitrogen on carbon storage, we have to ask the model where the additional nitrogen is distributed over time. Because the C/N of the plants and of the soil are fixed constants in the model, the changes in the carbon contents of plants and soil are simple multiples of the changes in their nitrogen contents. We will first explore this concept analytically by examining the steady-state solution to the model and asking the question, In a new steady state with added nitrogen, where does the added nitrogen reside? By studying the analytical properties of the model, we can determine how the different parameters affect the ultimate fate of added nitrogen. Then we will insert reasonable numerical values for the parameters and get a quantitative estimate of the effect over time of added nitrogen on stored carbon.

Denoting the steady-state solutions to Eqs. 2 through 4 with subscript zeroes (e.g., I_0, P_0, and $S_{N,0}$), we get from Eq. 2:

$$I_0 = \frac{f}{r} P_0 \tag{6}$$

and from Eq. 4:

$$S_{N,0} = \frac{r}{b} PI_0 = \frac{f}{b} {P_0}^2. \tag{7}$$

Eq. 3 provides no new information because of Eq. 5.

Suppose the system initially contains T_0 units of nitrogen and this amount is augmented by ΔT to a new value T_1:

$$T_1 = T_0 + \Delta T. \tag{8}$$

Hence,

$$\Delta P + \Delta I + \Delta S_N = \Delta T. \tag{9}$$

Using Eqs. 6 and 7, this can be rewritten as

$$\Delta P + \frac{f}{r}\Delta P + \frac{f}{b}\Delta(P^2) = \Delta T. \tag{10}$$

If the amount of added nitrogen is relatively small compared with the amount initially in the ecosystem, then we can replace $\Delta(P^2)$ with $2P_0\Delta P$ and obtain:

$$\Delta P = \frac{\Delta T}{1 + (f/r) + 2(f/b)P_0}. \tag{11}$$

Now let's insert some reasonable numerical values for the parameters. We will use units of $t(N)/km^2$ for the stocks (P, I, S_N) where t(N) refers to tons of nitrogen and t(N)/year for the flows. For forests, reasonable values for the initial stocks are $P_0 = 200$, $I = 1$, and $S_{N,0} = 800$. Very roughly, 20% to 40% of the nitrogen in trees is found in foliage, and much of this foliar nitrogen falls to the ground each year. So let's assume that about one-fourth of the nitrogen in the trees is returned to the ground each year. Hence, $f{P_0}^2/P_0 \sim 1/4$ or $f \sim (1/4)/200 = 0.00125$, with units of $[t(N)]^{-1}$ $(\text{year})^{-1}$. This, in turn, implies that $r \sim fP_0/I_0 \sim 0.25$ with the same units, and $b \sim f{P_0}^2/S_{N,0} \sim 0.0625$.

Inserting these values into Eq. 11, we get:

$$\Delta P = \frac{\Delta T}{1 + 0.005 + 8} = 0.111\Delta T. \tag{12}$$

In summary, only about 11% of the added nitrogen will ultimately reside in the trees. Using Eqs. 6 and 7, we can readily show that the amount of added nitrogen ultimately residing in the inorganic pool (I) is an even smaller percentage of ΔT, and that most of the added nitrogen (~89%) will end up in the form of soil organic matter (Exercise 2). Because the C/N of soil organic matter is ~25, not 300, this result casts doubt on the idea that nitrogen deposition will cause the sequestration of huge amounts of carbon.

Using our numerical values for the parameters in Model 1, we can also examine the time trajectories of the stock variables subsequent to the deposition of additional nitrogen to the system. In our first numerical "experiment," we start the system in steady state and then perturb it with a one-shot doubling of the value of I from 1 to 2. A spreadsheet approach allows us to readily determine the time dependence of P, I, and S_N following the addition of nitrogen to the system.

The first few lines of the spread-sheet calculation to determine the consequences of Eqs. 2 through 4 are shown below. We have assumed a time step of 0.01 years and the parameter values as given above.

	A	B	C	D	E	F	G
1	t	P	I	S_N	dP/dt	dI/dt	dS_N/dt
2	0	200	1 (or 2)	800	0.25*B2*C2 – 0.00125*B2*B2	0.0625*D2 – 0.25*B2*C2	0.00125*B2*B2 – 0.0625*D2
3	A2 + 0.01	B2 + E2*0.01	C2 + F2*0.01	D2 + G2*0.01	copy E2	copy F2	copy G2
4	copy A3	copy B3	copy C3	copy D3	copy E3	copy F3	copy G3
5	etc.	etc.	etc.	etc.	etc.	etc.	etc.

Figure IV-6 shows the numerical results for the case of an initial doubling of the value of I from 1 to 2. At the end of 10 years, the system is nearly in steady state, the plants have taken up ~11% of the added nitrogen, and the soil has taken up ~89%, as expected. About 0.05% of the added nitrogen resides in the inorganic pool (not shown in the figure). There is an initial sharp increase in plant nitrogen fol-

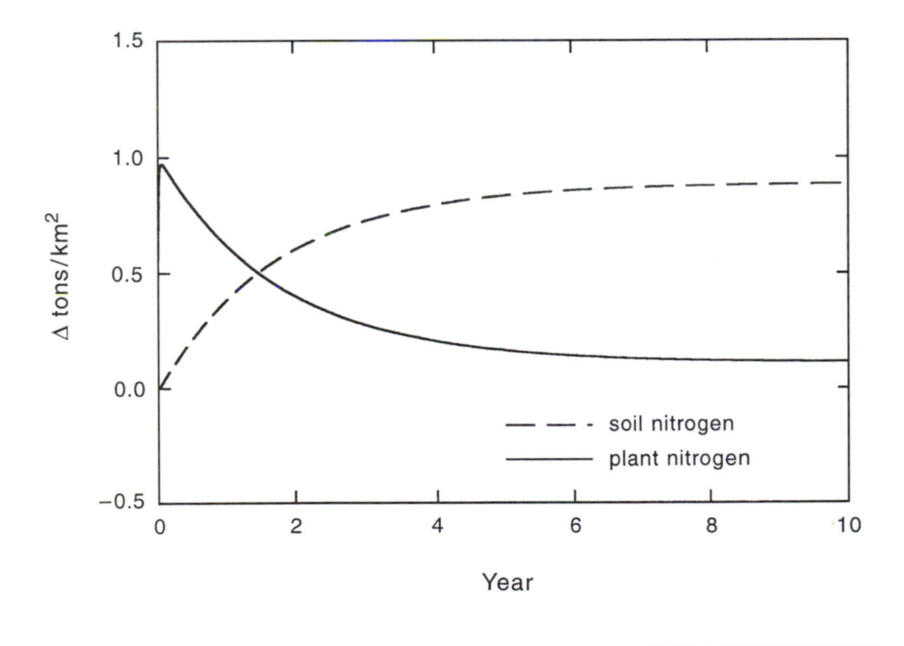

Figure IV-6 Results from Model 1: response of plant and soil nitrogen to an initial doubling of the amount of inorganic nitrogen from 1 to 2.

lowed by a rapid decline; that decline causes a corresponding increase in soil nitrogen, which rises and then approaches a new steady state.

We can also examine the effect of a continuing annual increase, by 1 t(N) per km^2 per year, in the amount of inorganic nitrogen in the soil. To do so, we alter the spread sheet by replacing the instructions in element C3 to: C2 + F2*0.01 + 0.01. The additional term, +0.01, means that at each time step, an extra quantity of nitrogen equal to 0.01 t(N) per km^2 is added. Because the time step is 0.01 year, this ensures that each year, 1 t(N) per km^2 is added. As seen in Figure IV-7, the buildup in the plant and soil nitrogen compartments is linear with time after an initial slow start in the buildup in soil nitrogen and an initial, more rapid, increase in plant nitrogen. In the linear phase, after about 5 years, the model tells us that the annual increases in soil and plant nitrogen, respectively, again correspond to about 89% and 11% of the annual nitrogen addition.

We can convert this to global carbon units by noting that the back-of-the-envelope calculation assumed that 100% of the added nitrogen ends up in wood with a C/N of 300. That calculation concluded that 1.5×10^9 tons(carbon)/year would be sequestered. However, with 89% of the nitrogen ending up in soil with a C/N of 20, and assuming that all the plant-sequestered nitrogen is stored in wood (an assumption we will examine later), Model 1 tells us that only $1.5 \times 10^9 \times [0.89(20) + 0.11(300)]/300 = 2.75 \times 10^8$ tons(carbon)/year

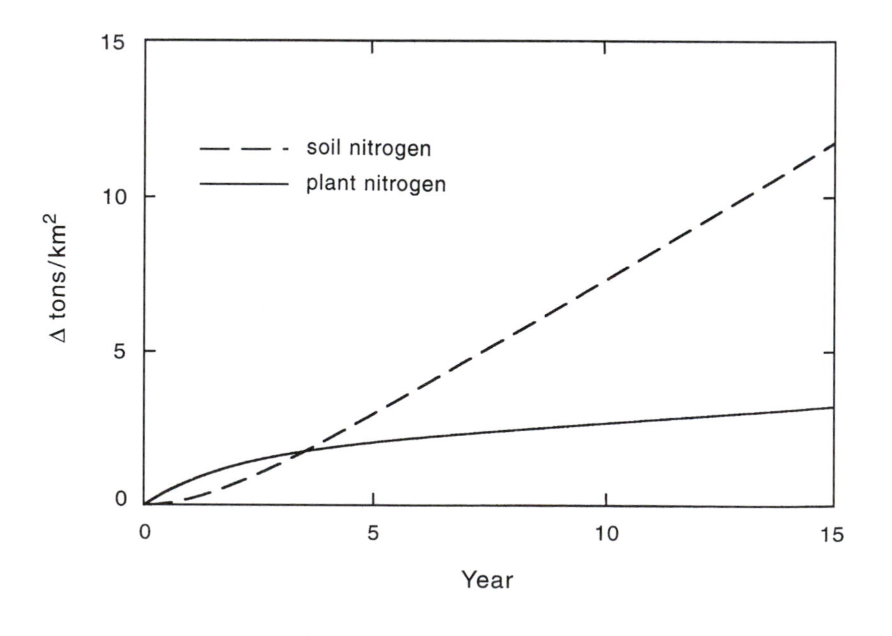

Figure IV-7 Results from Model 1: response of plant and soil nitrogen to a steady incease, by 1 t(N)/km^2/year in the amount of inorganic nitrogen.

will be sequestered. Although this amount is not negligible, it is too small to account for the missing carbon.

Model 2. To improve on our core model, we relax approximations 1 and 2 from our list above. We do this by adding explicit variables for soil carbon (S_C) and for the nitrogen content of the microbial population (M). By allowing the carbon content of soil to vary, we are also allowing the C/N of soil to change in response to nitrogen deposition. We do not allow for the possibility of a varying C/N in plants or microorganisms, and thus we maintain approximation 3 (along with approximations 4 through 12).

We continue to assume that the carbon dioxide content of the atmosphere is unchanged by the additional carbon uptake accompanying nitrogen deposition (i.e., approximation 5), but in Exercises 8 and 9 you will have a chance to explore the consequences of relaxing this assumption. Although the nitrogen cycle will remain closed in Model 2, the carbon cycle is not closed because we do not include inorganic carbon (e.g., the carbon dioxide in the atmosphere) explicitly in the model, and thus the sum of the remaining pools of carbon need not be constant.

A set of equations that corresponds to all these assumptions is as follows:

$$\frac{dP}{dt} = rPI - fP^2 \tag{13}$$

$$\frac{dM}{dt} = aMIS_C - gM^2 \tag{14}$$

$$\frac{dI}{dt} = bMS_NS_C - rPI - aMIS_C \tag{15}$$

$$\frac{dS_N}{dt} = fP^2 + gM^2 - bMS_NS_C \tag{16}$$

$$\frac{dS_C}{dt} = hP^2 + jM^2 - wMS_C. \tag{17}$$

In these equations, the terms have the following meanings:

fP^2 and hP^2 are the losses by death of plant nitrogen and carbon, respectively;

gM^2 and jM^2 are the losses by death of microorganism nitrogen and carbon, respectively;

rPI is the rate of uptake of inorganic nitrogen by plants (plant nitrogen assimilation);

$aMIS_C$ is the rate of uptake of inorganic nitrogen by microorganisms (microbial nitrogen immobilization);

bMS_NS_C is the rate of inorganic nitrogen formation resulting from microbial processing of organic matter (gross nitrogen mineralization).[25]

It follows from Eqs. 13 through 16 that $d[P + I + M + S_N]/dt$ is identically equal to 0, satisfying the constraint that nitrogen is conserved. Note that we have again taken the death rates for plants and also for microorganisms to be density-dependent and, for simplicity, have dropped terms corresponding to linear death rates, leaving us only with the terms proportional to P^2 and M^2.

The ratio h/f is the value of C/N for fresh litter fall. For some ecosystems, h/f would also be C/N of the living plants—for example, in an annual grassland, in which each year's crop of living plant matter is added as fresh litter to the soil each year. But for a forest, $(\mathrm{C/N})_{\text{plant litter fall}} \neq (\mathrm{C/N})_{\text{living trees}}$ because the parts of trees that die each year and are added to the soil have a C/N value that is considerably smaller than that of the whole tree. In general, we could write

$$\frac{h}{f} = \left(\frac{\mathrm{C}}{\mathrm{N}}\right)_{\text{plant litter fall}} = \frac{\sum_{i=1}^{n} \mathrm{C}_i/\tau_i}{\sum_{i=1}^{n} \mathrm{N}_i/\tau_i'}$$

where the C_i and N_i are the carbon and nitrogen contents of the n compartments that comprise the tree (e.g., leaves, twigs, trunk, fine roots, coarse roots) and the τ_i and τ_i' are the turnover times for carbon and nitrogen in these compartments. In contrast to the plants, for which annual litter fall does not necessarily have the same C/N as does the standing stock of live biomass, the C/N of dying microorganisms (j/g) is practically the same as that of the living ones.

To analyze the consequence of nitrogen deposition to this system, we proceed as we did with Model 1, finding the relationships among the changes in the steady-state solutions corresponding to a change in T. The steady-state solutions to the model are given by setting each time derivative equal to zero, thereby determining the constant values P_0, M_0, $S_{N,0}$, S_C, and I_0. After some messy algebra, the following rela-

25. The "net nitrogen mineralization rate" is the difference between the gross nitrogen mineralization rate and the microbial nitrogen immobilization rate. Thus, the net nitrogen mineralization rate $= bMS_NS_C - aMIS_C$. In steady state, it equals the plant nitrogen assimilation rate, or rPI. The trilinear functional forms we have chosen to describe the nitrogen immobilization rate and the gross mineralization rate may seem somewhat arbitrary and overly complicated. The use of these forms is motivated by the observation that immobilization would cease in the absence of any of the three quantities M, I, or S_C, and gross mineralization would cease in the absence of any of the quantities M, S_N, or S_C.

tionships can be derived:

$$\Delta S_N = \frac{af}{rb}\left[1 - \frac{jf}{hg}\right]\Delta P \tag{18}$$

and

$$\Delta S_C \cong \frac{hgr}{2wafS_{C,0}}\Delta P. \tag{19}$$

Eq. 18 is exact and valid for any magnitude of nitrogen deposition (that is any sized value of ΔT), but Eq. 19 is based on the approximation that $\Delta M \ll \Delta P$ and that $\Delta S_C \ll S_{C,0}$. Both these approximations are easy to justify: the equations that relate ΔM and ΔI to ΔP inform us that because the amounts of nitrogen stored in microbes and in the inorganic nitrogen pool are very small compared with P and S_N, the changes in those nitrogen pools are insignificant (Exercise 3).

For numerical values of the parameters, we retain the Model 1 values (and units) where appropriate, and thus $P_0 = 200$, $I_0 = 1$, and $S_{N,0} = 800$. A reasonable value for M_0 is 4. We express $S_{C,0}$ in units of $t(C)/km^2$ and assign it the value 20,000. As in Model 1, $r = 0.25$ and $f = 0.00125$. Assuming that the microbial populations turn over their nitrogen about twice per year, we get $gM_0{}^2 \sim M_0/(0.5y)$, or $g = 0.5$. Taking C/N of fresh plant litter to be 25, we get $h/f = 25$, or $h = 0.03125$. Taking C/N of microorganisms to be 8, we get $j/g = 8$, or $j = 4$. Hence, the parameter combination jf/hg in Eq. 18 has the value 0.32.

To determine the constant a, we note that in steady state, $aMIS_C = gM^2$, and so $a = gM/IS_C = (0.5)(4)/[(1)(20{,}000)] = 1 \times 10^{-4}$. To determine the constant b, we note that in steady state $bM_0S_{N,0}S_{C,0} = aM_0I_0S_{C,0} + rP_0I_0$, and hence $b = [(10^{-4})(4)(1)(20{,}000) + (0.25)(200)(1)]/[(4)(800)(20{,}000)] = 9.0625 \times 10^{-7}$. To determine the constant w, we use the steady-state condition $wMS_C = jM^2 + hP_0{}^2$, or $w = [(4)(4^2) + (0.03125)(200)^2]/[(4)(20{,}000)] = 0.016425$.

Substituting these values into Eqs. 18 and 19, we get

$$\Delta S_N = 0.38\Delta P \tag{20}$$

and

$$\Delta S_C \cong 48\Delta P. \tag{21}$$

Because nearly all the added nitrogen will eventually reside in either the plants or the soil organic pool (Exercise 2), Eq. 20 implies that ~70% of the added nitrogen ends up in the living trees and ~27% in the soil. Eqs. 20 and 21 imply that the soil will gain 48/0.38 = 105 t(C) for every t(N) that the soil gains.

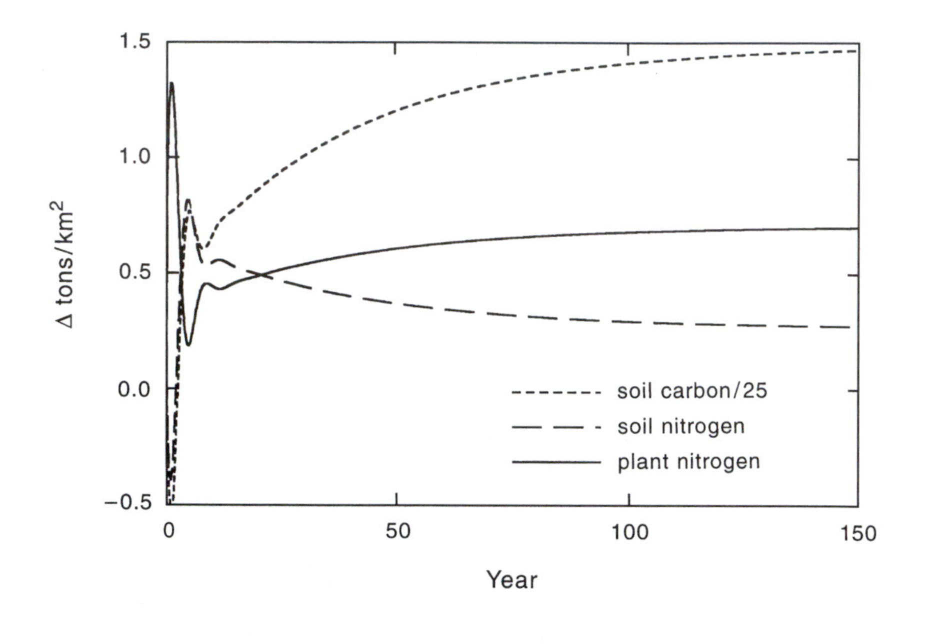

Figure IV-8 Results from Model 2: response of soil carbon and plant and soil nitrogen to an initial doubling of inorganic nitrogen from 1 to 2.

So, returning to the original calculation that motivated our model, what can we expect to see if human activity results in the deposition of 5×10^6 t(N)/year on forests? According to Eq. 20, ~70% of the added nitrogen each year will eventually reside in living trees, or $\sim 3.5 \times 10^6$ t(N). Using an average C/N of 230 for living trees yields a total of 8.1×10^8 tons of new carbon stored in trees each year. From Eq. 21, the soil carbon will increase by $48 \times 3.5 \times 10^6 = 1.7 \times 10^8$ t(C) each year.

Model 2 thus tells us that a total of $\sim 10^9$ tons of carbon will eventually be sequestered for each year's deposition of anthropogenic nitrogen, about two-thirds the value obtained from our simple back-of-the-envelope estimate at the start of this problem. The time dependence of carbon sequestering may be so slow, however, that our results may have little bearing on today's carbon budget. To explore this possibility, we turn to a numerical solution to the Eqs. 13 through 17, using the spread-sheet approach that we used for Model 1.

Figure IV-8 shows results for the case of an initial doubling of the value of I from 1 to 2. To fit ΔS_C on the same scale as ΔP and ΔS_N, I have also plotted $\Delta S_C/25$. In a reversal of the results from Model 1, at the end of 100 years, plants have taken up ~70% of the added nitrogen, and soil has taken up 27%. There is an initial sharp spike of plant nitrogen and a sharp transient loss of soil nitrogen and carbon; oscil-

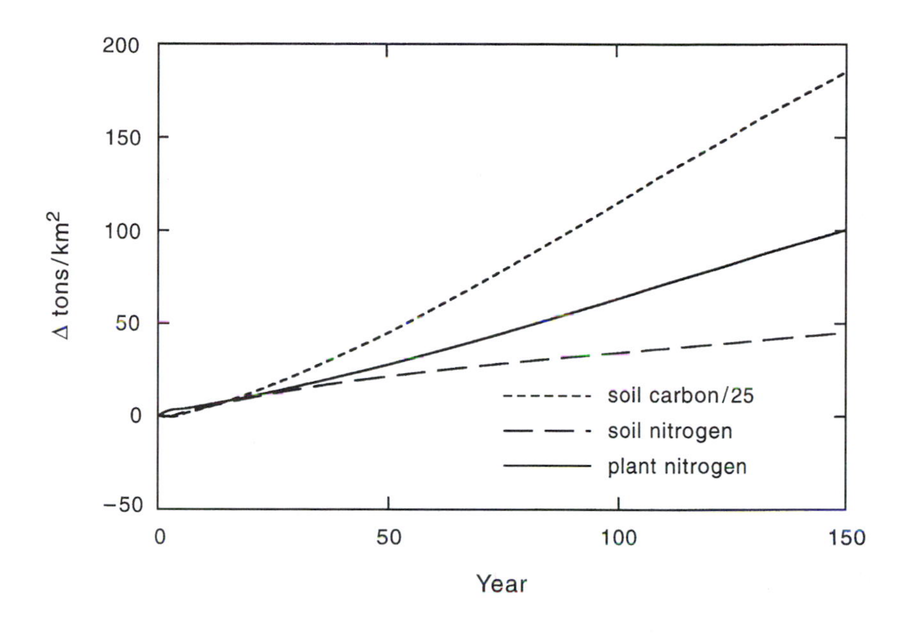

Figure IV-9 Results from Model 2: response of soil carbon and plant and soil nitrogen to a steady input of 1 $t(N)/km^2/year$.

lations occur that eventually damp out to the new steady state described by Eqs. 20 and 21.

If inorganic nitrogen is added steadily to the system, at 1 kg(nitrogen) per km^2 per year, the resulting time trajectories of plant and soil nitrogen and soil carbon predicted by Model 2 would be as shown in Figure IV-9. During the first 5 years, there is marked curvature in the time trajectories, followed by a more nearly linear increase with time (although some curvature persists out to 150 years). Eventually, in the linear phase, the annual increments in plant and soil nitrogen and soil carbon are compatible with Eqs. 20 and 21. Thus, Model 2 implies that a considerable amount of carbon can be sequestered as a result of anthropogenic nitrogen deposition, provided the nitrogen stored in trees is associated with carbon at the average C/N value for trees.

This result could, however, be misleading because the nitrogen added to P may not be associated with newly sequestered carbon in the ratio of $C/N = 230$ characteristic of the average tree prior to the added deposition. For example, suppose we think of a tree as having two nitrogen pools: the foliage with $C/N \sim 20$ and the remaining, woody, parts with $C/N \sim 300$, and the averaged $C/N \sim 230$. Moreover, assume that these C/N values do not change even when additional nitrogen is incorporated into the living trees. If initially one-fourth of all the nitrogen is in the foliage, but the newly deposited nitrogen is apportioned preferentially to the foliage, then the average C/N of the tree will be reduced. A similar shift in the C/N of soil is indeed predicted by Model 2, as is seen from Eqs. 20 and 21 (Exercise 4). To explore the sequestering of the added nitrogen among the different nitrogen pools in trees, we turn to Model 3.

Model 3. Here, approximation 3 in our list is relaxed. We do so by defining two plant variables, P_1 and P_2. P_1 is the nitrogen contents of the foliage, which we assume has a C/N of 20 and a turnover time of 1 year. P_2 is the nitrogen contents of the remainder of the tree, which has a C/N of 300 and a turnover time of 200 years. We also assume that nitrogen is taken up by trees and incorporated first into the P_1 compartment at a rate proportional to the product of I and $P = P_1 + P_2$; we assume that nitrogen is incorporated into the P_2 compartment from P_1 at a rate proportional to the product of P_1 and P_2.[26]

Model 3 is thus described by the following equations:

$$\frac{dP_1}{dt} = r(P_1 + P_2)I - f_1 P_1{}^2 - kP_1P_2 \tag{22}$$

$$\frac{dP_2}{dt} = kP_1P_2 - f_2 P_2{}^2 \tag{23}$$

$$\frac{dM}{dt} = aMIS_C - gM^2 \tag{24}$$

$$\frac{dI}{dt} = bMS_NS_C - r(P_1 + P_2)I - aMIS_C \tag{25}$$

$$\frac{dS_N}{dt} = f_1 P_1{}^2 + f_2 P_2{}^2 + gM^2 - bMS_NS_C \tag{26}$$

$$\frac{dS_C}{dt} = h_1 P_1{}^2 + h_2 P_2{}^2 + jM^2 - wMS_C. \tag{27}$$

For our initial conditions, we take $P_1 = 50$ and $P_2 = 150$, giving a total of $P_0 = P_1 + P_2 = 200$ as before. Note that the average C/N of the tree is given by $[50(20) + 150(300)]/200 = 230$, identical to what we assumed before. Other than dividing up the plants into two compartments, Model 3 is similar to Model 2 in form. As before, we take $M_0 = 4$, $I_0 = 1$, $S_{N,0} = 800$, $S_{C,0} = 20{,}000$, and we assume the same ini-

26. Both these assumptions are very crude, and it is well worth exploring the sensitivity of the model output to them (Exercise 11). Our assumption that the gross rate of nitrogen flow to the foliage is proportional to the total plant nitrogen contents, $P_1 + P_2$, is based on the plausible idea that the whole tree participates in nitrogen uptake and that the rate depends on total root mass, which, in turn, is roughly proportional to total tree mass. The assumption that the transfer of nitrogen from foliage to the rest of the tree is proportional to the product, P_1P_2, is based on the analogy with predator–prey interactions as modeled in Problem IV-4; in the limit of either P_1 *or* P_2 vanishing, the transfer rate must be zero.

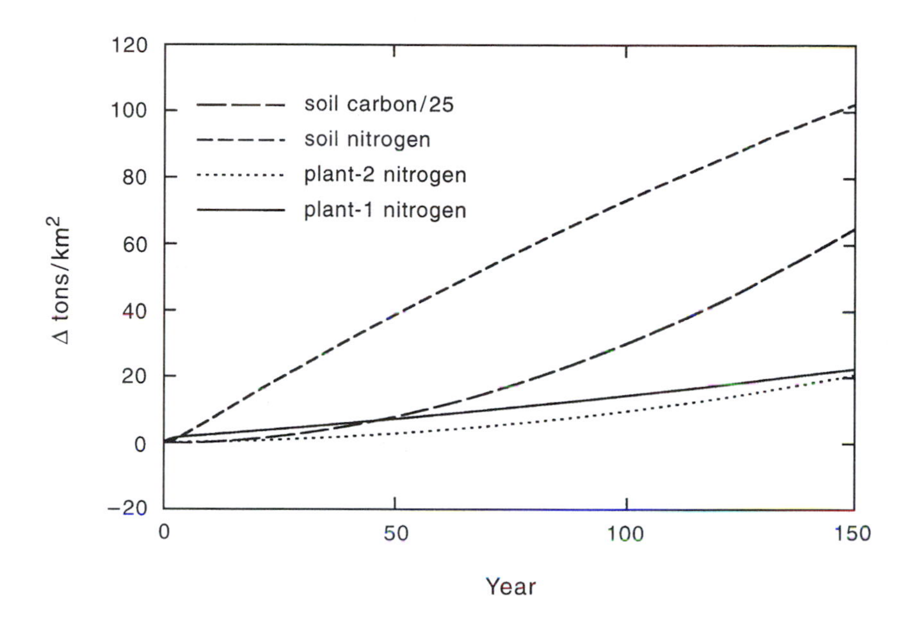

Figure IV-10 Results from Model 3: response of soil carbon, soil nitrogen, and plant nitrogen in both woody and leafy parts under a steady input of 1 t(N)/km^2/year.

tial rate of microbial nitrogen immobilization as in Model 2. These assumptions, along with the stated turnover times, allow us to determine the numerical values of the parameters from the steady-state solutions: $f_1 = 0.02$, $f_2 = 1/(3 \times 10^4)$, $r = 0.25375$, $k = 10^{-4}$, $h_1 = 0.4$, $h_2 = 0.01$, $g = 0.5$, $j = 4$, $a = 10^{-4}$, $b = 9.1796875 \times 10^{-7}$, $w = 0.0161125$.

A numerical solution to this model is given in Figure IV-10, for the case of a steady input of 1 t(N)/km^2 per year [in increments of 0.01 t(N)/km^2 per 0.01 year time step]. As shown, the behavior of this model is dramatically different from that of Model 2. At the end of 150 years, 68% of the added nitrogen is stored in soil. Only 14% is stored in the woody parts of trees, where the carbon-to-nitrogen ratio is very high, and thus much less total carbon is sequestered by the forest in this model.

We have arrived at an array of contradictory results about carbon sequestering—the answer depends on the level of detail in the model. In Model 1, the simplest, core model of the nitrogen cycle, most of the added nitrogen is predicted to end up in soil, where the relatively low C/N value implies only modest carbon sequestering. In Model 2, a much larger percentage of added nitrogen ends up in plants, and if the average plant value of C/N is assumed to remain constant, then a large amount of carbon is sequestered. In Model 3, in which distinct plant compartments correspond to foliage (with low C/N) and woody parts (with high C/N), most of the added nitrogen ends up in the soil

"I get to exercise and Bessie gets to recycle!"

(as in Model 1), and only about half of the nitrogen that ultimately resides in the trees ends up in the woody parts; thus, this model also predicts little carbon sequestering.

We conclude that modeling the effects of nitrogen deposition on carbon sequestering is a tricky business—modeler beware! Empirical studies do tend to support the prediction of Models 1 and 3. Only a small effect of nitrogen deposition on carbon sequestration in the forests of Europe and North America is observed.[27] Consistent with Model 3, the data suggest that ~70% of anthropogenically deposited nitrogen resides in soil.

EXERCISE 1: Redo the analysis leading to Eq. 11, assuming plant loss of nitrogen is linear in P, not quadratic. Then choose reasonable parameter values, and determine the percentage of added nitrogen that will ultimately reside in soil following an initial doubling of inorganic nitrogen (I).

EXERCISE 2: Show that in Model 1, the result that most of the added nitrogen ultimately resides in the soil, not in plants, is a consequence of our initial condition: $S_{N,0} \gg P_0, I_0$.

EXERCISE 3: (a) Verify Eqs. 18 and 19. (b) Show that the changes in M and I are very small compared with the changes in P, S_N, and S_C.

EXERCISE 4: What does Model 2 tell you about the effect of nitrogen deposition on the C/N ratio of soil? Can you explain why this counter-intuitive result is obtained from the model?

27. See Nadelhoffer, K., Emmett, B., Gunderson, P., Kjønaas, O., Koopmann, C., Schleppi, P., Tietma, A., Wright, R. 1999. Nitrogen deposition makes a minor contribution to carbon sequestration in temperate forests. *Nature* 398:145–48.

EXERCISE 5: In this exercise, you examine the consequences of our assumptions about the dependence of nitrogen turnover and microbial uptake on S_C in Model 2. In particular, you analyze the behavior of the model assuming soil carbon is no longer limiting turnover (in part a), and no longer limiting turnover or uptake (in part b).

(a) First, redo the derivation of Eqs. 18 and 19 for a revised Model 2 in which the term bMS_NS_C in the differential equations is replaced with $b'MS_N$, where $b' = 20{,}000b$, but all other parameters remain the same. What effect does this replacement have on the predicted percentage of nitrogen ultimately residing in soil?

(b) Now make a further replacement: let $aMIS_C$ also be replaced by $a'MI$, where $a' = 20{,}000a$, and repeat the analysis of the fate of added nitrogen.

What can you conclude about the sensitivity of the model output to our assumptions about soil carbon as a limiting factor?

EXERCISE 6: Often, models of biogeochemical cycles do not include a dynamic variable for the microbial population (that is, they make the second of our 12 listed assumptions), but they do include separate carbon and nitrogen variables and thus allow for time variation in the value of C/N of soil. Such models are intermediate in complexity between our Models 1 and 2. The justification often made for ignoring time variation in microbial population size is that only a relatively small amount of nitrogen and carbon is incorporated in the microbial population and thus variations in it do not matter very much. Here, you will examine this assumption. Construct a revised version of Model 2 in which every term, M, in the differential equations is replaced by the constant value $M = 4$, but let every other parameter remain unchanged. Then set up and run the spread sheet, and compare model output with the results in our Figures IV-7 and IV-8. Do microbes matter?

EXERCISE 7: Here, you examine the consequences of relaxing assumption 4. Add a term to Model 1 corresponding to a rate of loss of inorganic nitrogen from the system that is proportional to the difference between the actual value, I, and the steady-state value, 1: loss rate $= q(I - 1)$. Choose a value for q that corresponds to a loss of roughly half the added nitrogen in one year, and repeat the analysis of the ultimate fate of added nitrogen in Model 1 when I is initially doubled from 1 to 2 tons(N)/km^2.

EXERCISE 8: Here, you relax assumption 5 by constructing and analyzing a carbon-only model of plants, soil, microbes, and atmosphere in which carbon is conserved. Define new variables, P_C and M_C, that describe the carbon contents of plants and microorganisms. Your model will include P_C, M_C, S_C, and an atmospheric carbon pool, A. Write a set of differential equations analogous to our Model 2 for this closed carbon cycle, making the assumptions that microbial uptake of S_C occurs at a rate $w'M_CS_C$, that a fixed fraction of this carbon is incorporated into microbial biomass, and that the remainder is expelled to the atmosphere (A) as carbon dioxide. Assume further that plants incorporate carbon at a rate $r'P_CA$ and that the rates of loss of plants and microoganisms are quadratic functions of P_C and M_C, respectively. Make reasonable assumptions about the numerical values of the parameters and the initial conditions, and then work out the model prediction for the time-dependent distribution of carbon among P_C, M_C, S_C, and A if emissions of carbon dioxide to the atmosphere from fossil-fuel burning are increasing exponentially at a rate of 2.8%/year. (This exercise is an expansion of Exercise 7 of Problem IV-1, but now that you are armed with knowledge of biogeochemical modeling, you can approach it with considerably more skill.)

EXERCISE 9: Appropriately adjoin the model you developed in Exercise 8 with our Model 2, to form a combined carbon and nitrogen model. You need to think carefully about how to express the rates of input and output of C and N so that carbon and nitrogen conservation are automatically guaranteed by the form of the equations.

EXERCISE 10: Why does Model 3 predict a fate for added nitrogen that is so different from the prediction of Model 2?

EXERCISE 11: Explore the sensitivity of the output of our Model 3 to the form of the dependence on P_1 and P_2 of nitrogen uptake by P_1 and P_2. In particular, assume a dependence of plants on nitrogen that is of the same form as the dependence of foxes on rabbits given in Eq. 17 in Problem IV-4.

EXERCISE 12: Here, you have a chance to review what you learned about optimization in Chapter II and apply it to biogeochemistry. Our Model 2 (Eqs. 13–17) has the property that plants and microorganisms are both competitors (they compete for the common I pool) and mutualists (plants supply microbes with soil carbon, which the latter need as an energy source, and microbes supply plants with inorganic nitrogen, which they need as a fertilizer). Clearly, if microbes incorporate too much nitrogen for their own growth, plants will be starved for nitrogen, and ultimately the microbes will suffer for lack of plant-fixed carbon and their growth will be impaired. On the other hand, if the microbes take too little, their growth will also be suppressed. Suppose that the value of the parameter, b, governing the rate of gross nitrogen mineralization is fixed, but that microbes can evolve over time to optimize the value of the nitrogen uptake parameter, a. What ratio of a/b maximizes microbial biomass in the steady state? [*Note*: This is not a simple exercise; if you want to read more about this concept, see Harte, J., Kinzig, A. 1993. Mutualism and competition between plants and decomposers: Implications for nutrient allocation in ecosystems. *Am. Natural.* 14:829–46.]

Chapter V
Stability and Feedback

Gonzalo: "All trouble, torment, wonder and amazement inhabits here."

Introduction

From the fury of a tempest and the disruption it can cause to the mundane events of everyday existence, experiences with feedback and with stability or instability abound. Here are some everyday examples:

- The lid covering a pot of vigorously boiling water is pushed up by the steam pressure, thereby letting off steam, slowing the boil, and allowing the lid to plop back in place.
- Attempting to drive slowly and in reverse, the driver of a car pulling an attached trailer finds himself in a swaying motion of increasing amplitude.
- A job-seeker gets increasingly rattled and depressed over the course of an interview as she notices the questioner—a fatigued personnel manager not focusing clearly—seems confused by her answers to the questions.
- A drama group gives a stellar performance in response to intense audience enthusiasm.
- Voices at a cocktail party become increasingly loud so that conversations within each of the clusters of people can be heard.
- A small gully on a bare hillside deepens and widens as more runoff water is channeled down it.
- A dark stone placed on a fresh snowpack warms in the sun and melts the snow around it, causing a little bowl-shaped depression that focuses more sunlight and accelerates the melting.
- Because of favorable weather conditions, a population of insects grows to such high numbers that food supply becomes limiting and the population starts to decline.
- A population of fish in the sea is reduced by overfishing to such a low population density that males and females have trouble finding one another, thereby reducing the per-fish reproduction rate, and the population plummets further.

Stability means, in general terms, just what you think—resistance to a perturbing influence, or "steady as she goes." An unstable situation is one that is altered easily and dramatically by some force. Following this introduction, I will provide a mathematically precise definition of stability and instability.

In many circumstances, instability is worrisome. For example, if a small increase in the concentration of greenhouse gases, such as carbon dioxide and methane, in Earth's atmosphere results in a relatively small increase in global mean temperature, we might be able to live comfortably in the slightly altered climate. But if the Earth's climate system is unstable in the sense that a small amount of warming triggers further warming and eventually a sizeable portion of the polar ice caps melt, then the result could be catastrophic for humanity and many other species. Similarly, if the inadvertant introduction into a wilderness area of just a few individuals of an exotic species of plant results in the progeny of those individuals eventually dominating and displacing the native plant populations and causing extinctions, then that botanical instability would be widely considered an ecological catastrophe.

Both examples suggest the possibility that a system may be stable against a small initial disturbance but unstable if the initial disturbance exceeds a certain threshold. For instance, if only two or three individuals of an exotic plant species are introduced, then the chance that the species will expand in population may be negligible, but if several dozen individuals are introduced, then the chance could be high.

Common to all the examples in our list above is either reinforcement or inhibition. *Reinforcement* means that one event triggers a response that amplifies that event. Reinforcing processes are examples of positive feedback. Note that there need not be anything *positive*, in the everyday use of that word, about the sequence of events. The job-seeker certainly did not feel positive about the interview, and soil erosion is not a positive phenomenon.

There is also negative feedback, in which the initiating event triggers a response elsewhere in the system that inhibits or dampens that event. For instance, the air in a partly deflated tire on a fast-moving car will heat up more than normal because of increased internal friction in the wobbly rubber; at higher temperature, the air pressure in the tire increases, thereby causing the tire to stiffen and reducing the internal friction. (Please do not conclude that speedy driving is a sensible response to a deflated tire!)

Although the examples above mostly involve just two components—for example, the interviewer and the interviewee—sometimes feedbacks operate through a chain of indirect intermediaries. Although we did not treat the topic discussed in Problem II-5 as a feedback situation, it in fact is. Figure II-5 from Problem II-5, which illustrates the interactions among ecosystem health, economic activity, and human welfare, shows how feedback can involve more than just two interacting components.

Sometimes the assignment of a sign for a feedback process may seem ambiguous. For example, if the job-seeker had come to the interview full of enthusiasm, then the fatigued interviewer would exert a dampening, not a reinforcing, influence on the initial state of the interviewee. Does that mean the feedback was negative? No. To understand whether an influence is positive or negative, we need to ask whether the changes in the state of the system components are reinforced or dampened by the interaction between components. What makes the feedback positive is that the change (from enthusiasm to being rattled) was reinforced by the interaction. When a confused interviewer rattles a job-seeker, thereby causing further confusion, the direction of change is reinforcing. By itself, the initial turnabout in the mood of the job-seeker from a state of enthusiasm to a state of despair tells us nothing about the sign of the feedback.

Feedbacks can cause a system to be resistant to stress, or they can result in instability. They can result in predictable cyclic behavior or unpredictable, chaotic behavior. They can dampen temporal variation, leading to a steady state. As we shall see, there is a connection between the sign and magnitude of feedbacks and the stability or instability of a system. The connection is not as simple as you may, perhaps, be led to believe from the examples given above. We shall see, for example, that positive feedback does not necessarily mean that an instability occurs.

How can we predict if a system is likely to be unstable? What characteristics of a system tend to render it stable or unstable? How can we determine if and where a threshold exists, below which a system is stable and above which it is not? These are among the questions we will explore and provide tools for answering in this chapter.

EXERCISE: Which of the situations in the list above exemplify positive feedback, and which exemplify negative feedback?

Background: Stability

In all the sciences, the most widely used method for studying stability is applicable only to the study of initial disturbances that are small—so small, in fact, that we can pretend they are infinitesimal. We will start with that approach but talk about others later.

The method is useful only to the extent that we have a certain type of plausible mathematical description (a model) of the workings of the system. This description must be in the form of differential equations that express the rate of change of the system components. To illustrate the requisite type of model description of a system, let the components be labeled by a set of variables X_i. For example, X_1 might represent the global annual-averaged surface temperature, and X_2 might represent the areal extent of the north polar ice cap. The differential equations we seek would express the dependence of the time rate of change of each of the X variables on all of the X variables. Thus, the differential equations will be of the form

$$\begin{aligned}
\frac{dX_1}{dt} &= F_1(X_1, X_2, \ldots, X_N; a_1, a_2, \ldots, a_M) \\
\frac{dX_2}{dt} &= F_2(X_1, X_2, \ldots, X_N; a_1, a_2, \ldots, a_M) \\
&\cdot \\
&\cdot \\
&\cdot \\
\frac{dX_N}{dt} &= F_N(X_1, X_2, \ldots, X_N; a_1, a_2, \ldots, a_M).
\end{aligned} \tag{1}$$

The subscripts $(1, 2, \ldots, N)$ on the F functions tell us that these functions, which describe how the time derivatives of the X variables depend on the X variables themselves, may differ for each system variable. The a_i values are constant parameters that describe characteristics of the system. For instance, in the example of temperature and ice cover, one of the a parameters might be the melting point of ice.

A complication can result if some of the a parameters depend explicitly on time. For example, one of the a parameters might refer to the concentration of greenhouse gas in the atmosphere. Over short time periods, we can pretend that the concentration is constant and examine the coupling of ice cover and temperature. But if the concentration increases over time as a result of fossil-fuel combustion, then we may want to include in the model above a specified time-varying expression for a.

Alternatively, we may know that the greenhouse gas concentration varies over time, but we may need the model itself to tell us just how it varies. For example, if warming causes the release of carbon dioxide from the oceans, then the concentration is said to be dynamically coupled to temperature. Rather than using a prespecified expression for the time-dependence of the carbon dioxide concentration in our example above, we would introduce a new variable (X_3) to describe it. The time rate of change of X_3 would be a function of temperature, X_1.

In the following discussion, we assume that all time-varying parameters are represented as dynamic variables (that is, they are included among the X variables), so that the a parameters in Eqs. 1 are all constants. We also assume that in the absence of a disturbance to the system, all the variables (the X variables) are in a steady state, meaning that they do not depend on time. Given these assumptions and one more that I will explain shortly, we can derive explicit criteria that tell us whether a small disturbance to the system (an initial change in one or more of the X variables) damps out over time, in which case we say the system is stable, or grows over time, in which case we say the system is unstable. Although this is not the only way to define stability, it is a widely used one in environmental science.

We denote the original steady-state solution to the Eqs. 1 by the set of constants $\{X_{i,0}\}$. Because the time derivatives of the $X_{i,0}$ vanish, we have

$$F_i(X_{1,0}, X_{2,0}, \ldots, X_{N,0}; a_1, a_2, \ldots, a_M) = 0 \tag{2}$$

for $i = 1, 2, \ldots, N$.

We denote the initial disturbance, at time $t = 0$, to the $X_{i,0}$ by $\Delta X_{i,0}$. The choice of $t = 0$ for the time the disturbance is initiated is arbitrary and will not affect our results about stability. Practically speaking, in a complex system with many variables (i.e., a large value of N) the initial disturbance will directly affect only a small subset of the variables, so for most values of i, the $\Delta X_{i,0}$ will be zero; the formula for a stability criterion that we seek here, however, is quite general. To determine the behavior of the system subsequent to the onset of disturbance, we let $X_i(t)$ be the solution to the model for $t \geq 0$. Then we define new variables, $\Delta X_i(t)$, by

$$X_i(t) = X_{i,0} + \Delta X_i(t). \tag{3}$$

The $\Delta X_i(t)$ variables tell us how the X_i variables vary with time as they depart from steady-state values because of the disturbance. Note that, by the way, we have defined $\Delta X_i(t)$ and $\Delta X_{i,0}$, $\Delta X_i(t = 0) = \Delta X_{i,0}$. For those X_i variables that were not initially disturbed $[\Delta X_i(t = 0) = 0]$, the $\Delta X_i(t)$ variables may nevertheless be nonzero for $t > 0$ because the disturbance can spread from the variables that are initially disturbed to those that are not. The F_i functions describe the interactions among the variables that can lead to such a spreading of the disturbance.

The next step is to substitute the expressions in Eqs. 3 into the dynamic equations (Eqs. 1). To avoid clutter, I will not keep including the a parameters in our expressions for the F functions, but it is understood that the F functions do depend on these parameters. After substitution, we obtain

$$\frac{d[X_{i,0} + \Delta X_i(t)]}{dt} = F_i[X_{1,0} + \Delta X_1(t), X_{2,0} + \Delta X_2(t), \ldots, X_{N,0} + \Delta X_N(t)]. \tag{4}$$

Note that the left side of this equation can be replaced by

$$\frac{d[X_{i,0} + \Delta X_i(t)]}{dt} = \frac{d[\Delta X_i(t)]}{dt} \tag{5}$$

because the $X_{i,0}$ variables are time-independent.

At this stage, we make a critical assumption: **following the disturbance, the $\Delta X_i(t)$ variables remain small in comparison to the $X_{i,0}$ for all times subsequent to $t = 0$.** To see the importance of this assumption, let us back up a minute and examine a simpler situation. Consider a function $f(x)$. If we let $X = X_0 + \Delta X$, where ΔX is always small compared with X, then $f(X) = f(X_0 + \Delta X)$ can be approximated by the following expression:

$$f(X) = f(X_0 + \Delta X) \cong f(X_0) + \Delta X \frac{df(X)}{dX}\bigg|_{X=X_0}. \tag{6}$$

The symbol $[df(X)/dX]|_{X=X_0}$ simply means that the derivative $df(X)/dX$ is to be evaluated at $X = X_0$. The symbol $\cong$ means "approximately equal."

Eq. 6 is called a Taylor series approximation[28] to the function $f(X)$. It is a good approximation only if the second term on the right side is small compared with the first. When ΔX is small, the corrections to this approximation go to zero like $(\Delta X)^2$, and thus the approximation is better the smaller the value of ΔX is. If f is a function of two variables, $X_1 = X_{1,0} + \Delta X_1$ and $X_2 = X_{2,0} + \Delta X_2$, then we use the multivariable Taylor series approximation:

$$f(X_1, X_2) = f(X_{1,0}, X_{2,0}) + \Delta X_1 \frac{\partial f(X_1, X_2)}{\partial X_1}\bigg|_{X_1=X_{1,0};\, X_2=X_{2,0}} + \Delta X_2 \frac{\partial f(X_1, X_2)}{\partial X_2}\bigg|_{X_1=X_{1,0};\, X_2=X_{2,0}}. \tag{7}$$

The symbol $\partial f(X_1, X_2)/\partial X_1$ refers to the partial derivative of f with respect to X_1, meaning that the derivative is taken at fixed value of X_2.

Applying the multivariable Taylor series approximation to the right side of Eqs. 4, we can write

$$F_i[X_{1,0} + \Delta X_1(t), X_{2,0} + \Delta X_2(t), \ldots, X_{N,0} + \Delta X_N(t)] \cong F_i(X_{1,0}, X_{2,0}, \ldots, X_{N,0}) + \sum_{j=1}^{N} \Delta X_j \frac{\partial F_i(X_1, X_2, \ldots, X_N)}{\partial X_j}\bigg|_{X_1=X_{1,0};\, X_2=X_{2,0};\ldots,\, X_N=X_{N,0}}. \tag{8}$$

Combining Eqs. 2, 4, 5, and 8 gives us

$$\frac{d[\Delta X_i(t)]}{dt} = \sum_{j=1}^{N} \Delta X_j \frac{\partial F_i(X_1, X_2, \ldots, X_N)}{\partial X_j}\bigg|_{X_1=X_{1,0};\, X_2=X_{2,0};\ldots,\, X_N=X_{N,0}}. \tag{9}$$

---

28. See Appendix for the general form of the Taylor series expansion, of which the approximation in Eq. 7 contains just the two leading terms.

Note that we have inserted an equals sign into this equation. Strictly speaking, it should have an approximately equals sign, but because we will be solving Eq. 9 exactly later, it makes more sense to call it an exact equation from here on.

It is helpful to make mathematical notation as streamlined as possible—if nothing else, the equations are easier to read. Streamlining may even lead to shorter, and therefore cheaper, textbooks! If we define the matrix **A**, with matrix elements A_{ij}, by

$$A_{ij} = \left. \frac{\partial F_i(X_1, X_2, \ldots, X_N)}{\partial X_j} \right|_{X_1 = X_{1,0};\, X_2 = X_{2,0};\ldots,\, X_N = X_{N,0}} \tag{10}$$

and the vector $\mathbf{\Delta X}$ by

$$\mathbf{\Delta X} = \begin{matrix} \Delta X_1 \\ \Delta X_2 \\ \cdot \\ \cdot \\ \cdot \\ \Delta X_N \end{matrix}, \tag{11}$$

then Eqs. 9 can be rewritten in the more concise matrix-vector form[29]

$$\frac{d(\mathbf{\Delta X})}{dt} = \mathbf{A} \cdot \mathbf{\Delta X}. \tag{12}$$

These equations (9 and 12) state that when a system is disturbed, the displacement of each of its components changes over time at a rate that depends in a specified way on the displaced values of all the components.

As with any differential equation, the solution to Eq. 12 must satisfy a set of initial or boundary conditions. In our case, the initial conditions are that the $\Delta X_i(t = 0)$ equal the specified initial displacements $\Delta X_{i,0}$. Because the derivatives of the F_i that comprise the matrix elements of **A** (defined by Eqs. 10) are evaluated at the steady-state values of the X_i (the $X_{i,0}$), the matrix **A** is a constant matrix. Thus, the collection of N equations (Eq. 12) that describe the family of relationships among the ΔX_i is a coupled set of linear differential equations with time-independent coefficients.

Sets of equations of the form of Eq. 12 crop up frequently in various sciences. The equations are so important that people in various disciplines give the matrix **A** special names. Ecologists call it the "community matrix" because it describes how a community of species is

---

29. For a brief exposure to the rules governing matrix and vector algebra, see the Appendix, but for a more in-depth treatment of this topic, see C. Swartz's *Used math*, listed in Further Reading after the Appendix.

hooked together. For example, if $i = 1$ stands for rabbits and $i = 2$ stands for foxes, then $A_{12} = \partial F_{\text{rabbits}} / \partial X_{\text{foxes}}$ tells us how the rate of change of rabbits is directly affected by the foxes. As you might expect, A_{12} would be negative (the direct effect of more foxes is to reduce the number of rabbits); the magnitude of A_{12} is thus related to the predation rate of foxes on rabbits. [*Reader*: What is the meaning and sign of A_{21} in a rabbit–fox model?] Through the rest of this book, I shall adopt ecologists' terminology and refer to the matrix **A** as the community matrix, even if the system described by **A** is not made up of species of organisms.

Even though the equations stemming from Eq. 12 are complex because they interconnect N components, they possess a unique and exact analytical solution:

$$\Delta X_i(t) = \sum_{j=1}^{N} \sum_{\sigma=1}^{N} (\mathbf{C})_{i\sigma} (\mathbf{C}^{-1})_{\sigma j} \Delta X_{j,0} e^{\lambda_\sigma t}. \tag{13}$$

The λ_σ in these expressions are complex numbers[30] called eigenvalues. The matrix, **C**, is made up of columns of numbers called eigenvectors. Each of the N eigenvectors, $\mathbf{V}_\sigma$, corresponds to one of the N eigenvalues, λ_α, and satisfies the equation

$$\mathbf{A}\mathbf{V}_\sigma = \lambda_\sigma \mathbf{V}_\sigma. \tag{14}$$

The eigenvectors are independent of time because the matrix elements of **A** are. As can be seen from the way the eigenvalues multiply the time variable, t, in the exponents in the double summation in Eq. 13, the eigenvalues determine the time dependence of the $\Delta X_i(t)$. In particular, Eq. 13 implies that the ΔX_i behave like a linear combination of the $e^{\lambda_\sigma t}$.

Eq. 14 leads to a useful formula for calculating the eigenvalues. To see how, we have to introduce two entities, the unit matrix and the concept of a determinant of a matrix. The unit matrix, **I**, plays a role similar to that of the number 1 in the ordinary number system. **I** is a square matrix with ones along the diagonal and zeroes everywhere else. The elements of the unit matrix can be expressed by the kronecker-delta symbol δ_{ij}, which has the value 1 if $i = j$ and 0 if $i \neq j$: $(\mathbf{I})_{ij} = \delta_{ij}$. The matrix has the property that for any matrix $\mathbf{A} : \mathbf{AI} = \mathbf{IA} = \mathbf{A}$. Using the unit matrix, Eq. 14 can be reexpressed as

$$(\mathbf{A} - \lambda \mathbf{I})\mathbf{V} = 0. \tag{15}$$

30. By "complex numbers" we mean here that they can have both a real and an imaginary part. For a good review of complex numbers and other essential mathematical background material, including eigenvalues and eigenvectors, see C. Swartz's *Used math* (see Further Reading).

Formally, this equation has a solution given by

$$\mathbf{V} = 0 \cdot (\mathbf{A} - \lambda \mathbf{I})^{-1}, \tag{16}$$

a peculiar expression indeed. But the inverse of any matrix, **B**, such as $\mathbf{B} = \mathbf{A} - \lambda\mathbf{I}$, can actually be calculated, which is where the concept of the determinant comes in. The general formula for the determinant of matrix **B** is given in the Appendix. It is an awkward formula to use in practice because it expresses the determinant of **B** in terms of a certain sum over the determinants of smaller matrices obtained by striking out single rows and columns of **B**. For a 2×2 matrix, the formula reduces to a simple expression: determinant $(\mathbf{B}) = B_{11}B_{22} - B_{12}B_{21}$. And for a 3×3 matrix, the formula is determinant $(\mathbf{B}) = B_{11}B_{22}B_{33} + B_{13}B_{32}B_{21} + B_{31}B_{12}B_{23} - (B_{31}B_{13}B_{22} + B_{11}B_{32}B_{23} + B_{33}B_{21}B_{12})$.

The matrix elements of the inverse of a matrix **B** depend on the determinant of **B**:

$$(\mathbf{B}^{-1})_{ij} = \frac{M_{ji}}{\text{determinant } (\mathbf{B})}, \tag{17}$$

where the M_{ji} are the so-called cofactors of the matrix **B**, which are also defined in the Appendix. For Eq. 16 to have a nonzero solution, the inverse of $\mathbf{A} - \lambda\mathbf{I}$ must be infinite, so that the zero in the numerator is cancelled. The M_{ji} in Eq. 17 are all finite, so that equation tells us that the inverse of $\mathbf{A} - \lambda\mathbf{I}$ is infinite if and only if

$$\text{determinant } (\mathbf{A} - \lambda\mathbf{I}) = 0. \tag{18}$$

Hence, Eq. 18 is the condition that determines the eigenvalues! Then, once the eigenvalues are determined, Eq. 14 determines the eigenvectors. A sample calculation of eigenvalues and eigenvectors is given in the Appendix.

Returning to Eq. 13, when $t = 0$, the sum over σ can be obtained easily because the product of a matrix and its inverse is the unit matrix, **I**. Hence, using the rules for matrix multiplication (see Appendix) at $t = 0$, we recover the identity $\Delta X_i(0) = \delta_{ij}\Delta X_{j,0} = \Delta X_{i,0}$.

When $t > 0$, the sum over σ is more complicated because the terms $e^{\lambda_\sigma t}$ no longer equal 1, and thus the sum is no longer simply a product of two matrices. Nevertheless, some very important information about the stability of the steady-state solutions, $X_{i,0}$, can be gleaned from Eqs. 13. In particular, consider the limit of large t. Depending on the values of the λ_σ, the terms $e^{\lambda_\sigma t}$ will either blow up to infinity, damp out to zero, or oscillate as t becomes large. To see why, we first note that each λ_σ, being the solution to a polynomial equation, can be a complex number, with a real part, $\lambda_{\sigma,\mathrm{R}}$ and an imaginary part $\lambda_{\sigma,\mathrm{I}}$, so that $\lambda_\sigma = \lambda_{\sigma,\mathrm{R}} + i\lambda_{\sigma,\mathrm{I}}$. If the real part is greater than 0, then $e^{\lambda_{\sigma,\mathrm{R}} t}$ grows infinitely large as $t \to \infty$. In that case, Eqs. 13 tell us that some of the

$\Delta X_i(t)$ may grow to infinity as $t \rightarrow \infty$. Which ones blow up like that depends on which of the $X_{i,0}$ are nonzero and the specific entries in the matrix of eigenvectors, **C**.

In words, $\Delta X_i(t) \rightarrow \infty$ means that the initial and small perturbation to the system ultimately propagates into an infinite response. That is just what we mean here by instability. So, if at least one eigenvalue of the community matrix has a positive real part, then the system described by that matrix is unstable.

If the real parts of all the eigenvalues are less than zero, then $\Delta X_i(t) \rightarrow 0$ as $t \rightarrow \infty$. That, as you guessed, is what we mean by stability: the values of all the perturbations (the ΔX) eventually damp out to zero.

Finally, if some or all of the eigenvalues have zero real parts and the remaining ones have negative real parts, then as $t \rightarrow \infty$, some of the $\Delta X_i(t)$ can remain finite. The reason is that $e^{0^* t} = 1$. This situation is called neutral stability; the initial perturbation neither shrinks to zero nor grows to infinity. Instead, it propagates in a bounded manner.

Examples of each of these cases are illustrated here (Figure V-1) for the simple case in which the system is a ball free to roll on a surface with friction. In each case, the initial steady-state location of the ball is

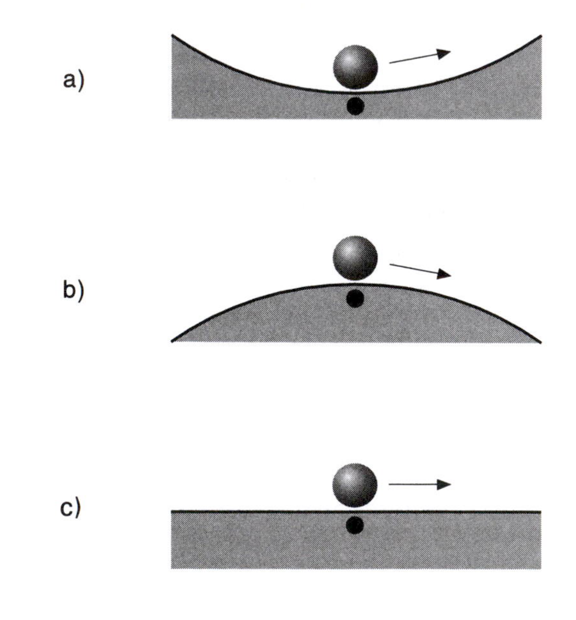

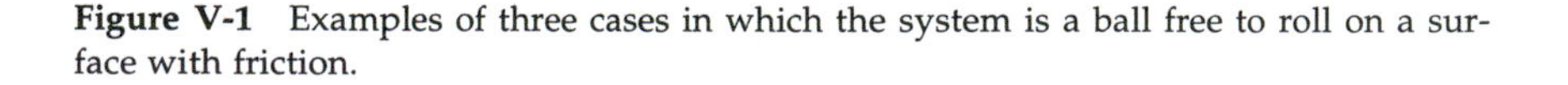

Figure V-1 Examples of three cases in which the system is a ball free to roll on a surface with friction.

denoted by the solid dot and an initial displacement from steady state is denoted by an arrow. In case *a*, the displaced ball will undergo a damped oscillation around the steady state, eventually returning to it. In case *b*, the displaced ball will depart further from steady state and never return to it. In case *c*, the ball neither returns to the steady state nor departs further from the point to which it was displaced. Case *a* depicts a stable configuration, case *b* depicts instability, and case *c* depicts neutral stability.

The attentive reader may be bothered by the idea of $\Delta X_i(t) \to \infty$. After all, the mathematical argument leading up to Eq. 13 was based upon the assumption that we could drop all but the first term in the Taylor series expansion, and this assumption required that the $\Delta X_i(t)$ were all relatively small. In fact, if any of the $\Delta X_i(t)$ grow too big, then our analysis is probably no longer valid, and Eq. 13 no longer is a good approximation. This does not mean a system with $\lambda_{\sigma,\mathrm{R}} > 0$ is in fact stable, but it does mean that $\Delta X_i(t)$ may not actually become infinite. The only situation where the analysis would be valid when the $\Delta X_i(t)$ grow large is when the F_i in Eqs. 1 are linear in the X_i, so that Eq. 9 is exact.

What about the imaginary parts of the eigenvalues—what do they tell us about the time dependence of the $\Delta X_i(t)$? Here, we use the fact (see Appendix) that $e^{i\lambda_{\sigma,\mathrm{I}}t} = \cos(\lambda_{\sigma,\mathrm{I}}t) + i\sin(\lambda_{\sigma,\mathrm{I}}t)$. Because these sine and cosine functions oscillate over time, the $\Delta X_i(t)$ will oscillate. Indeed, some of the many observed periodic or cyclic phenomena observed in nature can be predicted from examination of eigenvalues; if at least some contain nonzero imaginary parts, then the system will have oscillatory components.

If an eigenvalue has a nonzero imaginary part and a negative real part, then the combination of the ΔX_i values comprising that eigenvector will damp out in time, but in approaching zero they will oscillate around their destination. That is called damped oscillation. On the other hand, if the real part is positive, then the eigenvector will swing wildly between ever-larger positive and negative values as it approaches infinity. Thus, the eigenvalue corresponding to case *a* in Figure V-1 showing balls on surfaces must have had a nonzero imaginary part.

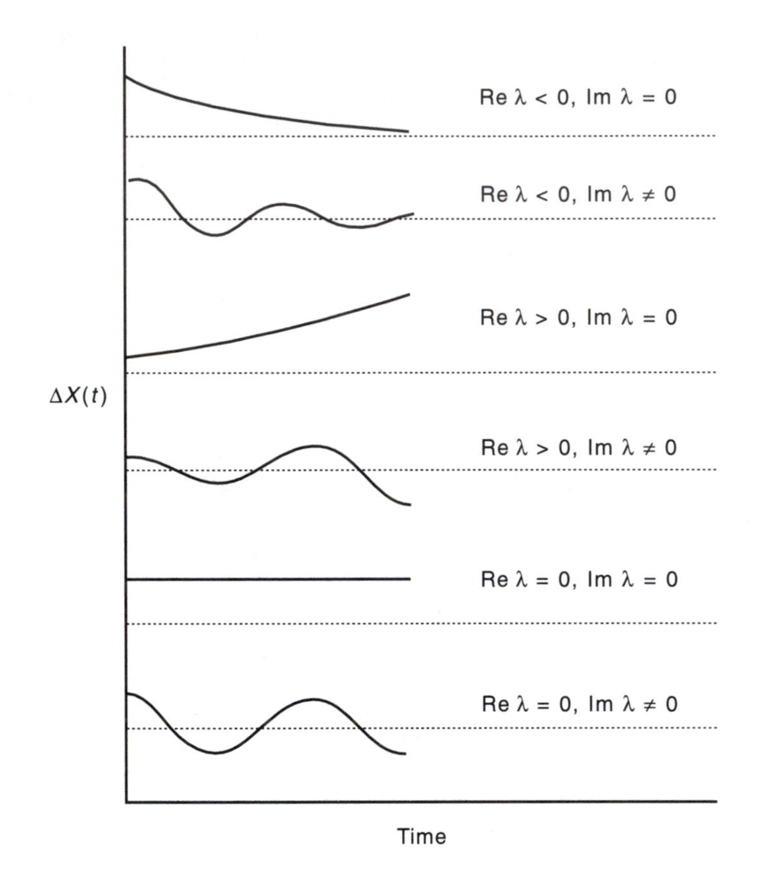

Figure V-2 How the eigenvalue affects the response of a 1-variable system to an initial perturbation. $\Delta X(t)$ is the time-dependent displacement of the variable from an initial steady state. If the real part of the eigenvalue is negative, the system is asymptotically stable, and $\Delta X(t)$ approaches zero (dotted line) as time $\to \infty$.

Background: Feedback

Our concern here is with the description of feedback and the development of quantitative tools with which we can explore the consequences of feedback. A useful way to picture a feedback process is to draw a set of labeled system components and connecting arrows that form one or more closed loops (see box). Each arrow is accompanied by a positive sign, negative sign, or zero, assigned according to the nature of the causal connection between components as shown in the box.

Picturing Feedback Processes

If an arrow extends from component A to component B and

> an increase in A causes an increase in B, then assign (+),
> an increase in A causes a decrease in B, then assign (−),
> an increase in A has no effect on B, then assign (0).

Consider the following example:

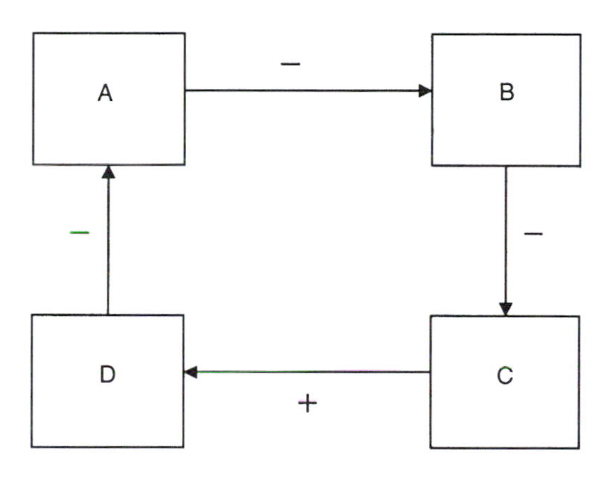

This feedback loop indicates a process in which an increase in A causes a decrease in B, an increase in B causes a decrease in C (or, equivalently, a decrease in B causes an increase in C), an increase in C causes an increase in D, and an increase in D causes a decrease in A. Because the product of (−)(−)(+)(−) is (−), this example describes a negative feedback loop. (An initial increase in A triggers a sequence of events that reduce the original increase.)

In general, the overall sign of the feedback process is given by the product of the signs of all the linkages around a closed loop. Note that each of the signs is determined solely by the nature of the linkage between the two linked components and is independent of the other signs in the sequence. In this example, component A might refer to surface temperature, B to soil moisture, C to area of desert, and D to surface albedo. If C referred to area of non-desert, then the signs linking B to C and C to D would be reversed, but the overall sign of the feedback would be unchanged.

To be more quantitative, we need to make some definitions. We are interested in the situation in which an initial change in a system occurs and is either enhanced or damped by feedback. In the absence of feedback, the total change in the system is just the initial change, and we call that the direct effect. Thus, with no feedback,

$$\text{Total Effect} = \text{Direct Effect}.$$

Now, suppose there is a feedback process. With one pass around the feedback loop, the direct effect will be enhanced or dampened by a factor g:

$$\text{Total Effect} = \text{Direct Effect} + (g)\ (\text{Direct Effect}).$$

If g is positive, then the feedback is positive (total effect greater than direct effect), whereas if g is negative, the feedback is negative.

In practice, there will be infinitely many passes around the feedback loop, and each one leads to another factor of g multiplying the input:

$$\begin{aligned} \text{Total Effect} &= (\text{Direct Effect})(1 + g + g^2 + g^3 + \cdots) \\ &= \frac{\text{Direct Effect}}{1 - g} \quad \text{if } g < 1. \end{aligned} \tag{19}$$

Now we can distinguish three ranges of value for g:

$$0 < g < 1 : \text{Total Effect} > \text{Direct Effect; positive feedback}$$

$$g < 0 : \text{Total Effect} < \text{Direct Effect; negative feedback}$$

$$g \geq 1 : \text{the sum } (1 + g + g^2 + g^3 + \cdots) \text{ no longer converges; that is,}$$

$$\text{Total Effect} = \infty. \text{ This corresponds to instability.}$$

Given a dynamic model that incorporates a feedback mechanism, it is not difficult to actually calculate the quantity g. Suppose, by way of example, that Earth's surface temperature, T_S, can be expressed as a function, F, of various factors, and that some of these factors themselves depend on T_S:

$$T_S = F[p_1(T_S), p_2(T_S), \ldots]. \tag{20}$$

For example, $p_1(T_S)$ might represent the albedo of Earth's surface, which would change if warming melts polar and glacial ice or induces changes in vegetation cover.

Then,

$$g = \sum_i \frac{\partial F}{\partial p_i} \frac{\partial p_i}{\partial T_S} = \sum_i g_i. \tag{21}$$

A derivation of Eq. 21 and a more complete derivation of Eq. 19 is presented in *COW-1* (pp. 181–83).

Consider again just a single term, $i = 1$, in the sum in Eq. 21, with p_1 being surface albedo. The first partial derivative in this first product measures the effect of a change in surface albedo on global temperature. Estimation of that derivative could be carried out with a model such as the climate model presented in *COW-1* (pp. 160–71). Multiplying that derivative is the term $\partial p_1/\partial T_S$, which expresses the effect of temperature change on surface albedo. Evaluation of this term requires knowledge of the extent to which warming will induce a change in ice, snow, and vegetation cover and the effect of those changes on surface albedo. The effect on albedo of a shift in surface cover is usually fairly easy to estimate because the albedos of many surface conditions are measured (see Climate Data Appendix in *COW-1*, p. 263). More difficult is knowing how much snow and ice will melt and how vegetation cover will change under climate warming. In Problem V-2, you will get some practice estimating the magnitude of this feedback and its contribution to global warming.

This formalism (Eqs. 19–21) achieves the goal of summing an infinite number of traverses around the feedback cycle: (climate change) $\rightarrow$ (ecosystem change) $\rightarrow$ (climate change) $\rightarrow \ldots$. It is valid, however, only if the relationship between the state of the ecosystem and the state of the climate is linear or the perturbation is small in the sense that first-order Taylor series expansions of the state of the climate and of the ecosystem around the unperturbed state are valid.

In more concrete terms, suppose that the climate variable of interest is globally averaged surface temperature, T_S, expressed in kelvins. The change in the value of T_S is small in the sense above, if the change in the value of T_S is a small fraction of T_S itself. Practically speaking, the formalism would be generally considered useful if this fraction is ≤ 0.1, corresponding to a maximum temperature change of roughly 30 K.

Conventional general circulation models incorporate several important feedback mechanisms. One is called the ice–albedo feedback: warming induced by elevated atmospheric greenhouse gas concentrations results in the melting of ice and snow cover, which leads to a lower value of the Earth's albedo. Because a lowered albedo enhances the original warming, this process is a positive feedback mechanism. A second feedback mechanism involves water vapor. The amount of water vapor, itself a greenhouse gas, that the atmosphere can hold at constant relative humidity increases with rising atmospheric temperature. Combined with evidence that relative, not absolute, humidity is approximately invariant under warming, this information leads to another positive feedback mechanism. A third, of more uncertain magnitude and even sign, involves clouds. Clouds can warm Earth's climate by trapping outgoing, radiated heat and outgoing sunlight reflected back to space from below the clouds. They can also cool the planetary surface by reflecting or absorbing incoming solar radiation. The net effect depends on the altitude and optical properties of the clouds and the albedo of Earth's surface below the clouds.

Together, these feedbacks to the direct effect on globally averaged surface temperature of increasing greenhouse gas concentrations result in a value of the gain, g, in the range of 0.4–0.78. Much of the spread in this range reflects the uncertainty in the cloud feedback. The *direct effect* of an increase in greenhouse gas levels to a level that is effectively equivalent to a doubling of the pre–industrial-revolution CO_2 level (the "$2 \times CO_2$ scenario") is calculated to be roughly 1°C (see *COW-1*, pp. 176–86). Multiplying this direct effect by the factor, f, given by

$$f = \frac{1}{1-g} \tag{22}$$

from Eq. 19, we find that the *total effect* of the greenhouse gas increase is an elevation of surface temperature by an amount ranging from 1°C/(1 – 0.4), or 1.6°C, to 1°C/(1 – 0.78) or 4.5°C. These estimates stand today as widely cited values for the effect of the doubling of CO_2 on surface temperature. Inclusion of the climatic effects of aerosols results in somewhat lower estimates, however, because over much of the Earth's surface the *direct effect* of the sum of aerosol and CO_2 doubling is less than the direct effect of CO_2 doubling alone.

We note that the value of g is given by a sum over all the individual feedbacks that could possibly contribute to the total effect. Those individual feedbacks in the sum that are positive add to the value of g, whereas negative feedbacks reduce the value of g. At the upper limit of uncertainty of the combined effect of the ice–albedo, water-vapor, and cloud feedbacks, where $g = 0.78$, only a very small relative increase in g would be needed to exert a huge effect on the increase in surface temperature. For example, if one additional feedback process adds a term in the sum that raises the upper limit on g from 0.78 to 0.9, the effect would push the upper limit of warming up from 4.5°C to a value of 1°C/(1 – 0.9), or 10°C. Yet, compared with the value of the other feedbacks, this is not a large additional process.

EXERCISE 1: Using Eq. 14, show that Eq. 13 does indeed yield a solution of Eq. 12.

Problems

V-1. Stability of the Core Models

Examine each of the core models in Chapter IV to determine whether their steady-state solutions are stable against small perturbations.

• • • • • • •

Before turning to the five core models, let's examine the Malthus model that describes exponential growth of a population.

The Malthus model. This model reads

$$\frac{dX}{dt} = rX. \tag{1}$$

The general solution to this equation is $X(t) = X(0)e^{rt}$. If $X(0) \neq 0$, the case is not a steady state, and so the eigenvalue method we presented above cannot be used. Thus, we will examine only the stability of the case $X(0) = 0$, allowing us to determine if the Malthus model is stable against a small perturbation from "extinction."

The community matrix, **A**, for this equation is a 1×1 matrix because there is only one equation. Our rule for determining the community matrix requires that we evaluate the derivatives of the right sides of the dynamic equations at the steady state. For Eq. 1,

$$\mathbf{A} = \left.\frac{d(rX)}{dX}\right|_{X=0} = r. \tag{2}$$

The single eigenvalue of this 1×1 matrix is given by setting determinant $(r - \lambda) = 0$, yielding

$$\lambda = r. \tag{3}$$

If $r > 0$, then this eigenvalue is a positive real number. The system is unstable. An infinitesimal deviation from $X = 0$ will grow without bound. Unlike diamonds, extinction is not forever in the Malthus model. This situation is biologically not very realistic; a population that is extinct generally will not recover when only a very small number of individuals are reintroduced. We will see that a similarly unrealistic situation occurs when the population is described by the density-dependent logistic equation (Problem IV-2). In Exercise 2, you will get a chance to figure out how to construct a population growth model that is stable at the steady-state value of $X = 0$.

The core of pollutant stock-and-flow models. Our core equation is of the form

$$\frac{dX}{dt} = a - bX. \tag{4}$$

The steady-state solution to this equation is given by $X_0 = a/b$, and so the community matrix is given by

$$\mathbf{A} = \frac{d(a - bX)}{dX}\bigg|_{X=a/b} = -b. \tag{5}$$

Hence, the eigenvalue of Eq. 4 is

$$\lambda = -b. \tag{6}$$

Because b is a positive real number (assuming that pollutant transport is donor-controlled, as discussed in Problem IV-1), our system is said to be "neighborhood asymptotically stable"; moreover, when the system is perturbed from its steady state, it will return to its previous steady state without the oscillations characteristic of a system with an eigenvalue possessing a nonzero imaginary part.

The core of population models. Our core model is of the form

$$\frac{dX}{dt} = rX\left(1 - \frac{X}{K}\right). \tag{7}$$

This equation has two steady-state solutions: $X_0 = K$ and $X_0 = 0$. Consider, first, the stability of the state $X_0 = K$. The community matrix, evaluted at this steady state, is

$$\mathbf{A} = \frac{d[rX(1 - X/K)]}{dX}\bigg|_{X=K} = r - \frac{2rX}{K}\bigg|_{X=K} = -r. \tag{8}$$

Hence, the eigenvalue is

$$\lambda_{X=K} = -r, \tag{9}$$

and Eq. 7 is stable at the steady state $X = K$. Evaluating Eq. 8 at the other steady-state value of X, $X_0 = 0$, yields the eigenvalue

$$\lambda_{X=0} = r. \tag{10}$$

This eigenvalue has a positive real part, and so, just as with the Malthus equation, the extinction state is unstable.

The core of climate models. Our core equation is of the form

$$\frac{dX}{dt} = c - bX^4, \tag{11}$$

where, from Problem IV-3, X = temperature, $c = (\Omega/4K)(1-a)$, and $b = \sigma/K$. The steady state is $X_0 = (c/b)^{1/4}$. By now, you are getting so good at this that you can derive the result:

$$\lambda = -4c^{3/4}b^{1/4}. \tag{12}$$

Using $X_0 = (c/b)^{1/4}$, this eigenvalue can also be written in the form

$$\lambda = -4bX_0{}^3. \tag{13}$$

Either way, the eigenvalue is real and negative, indicating stability of the steady-state solution.

The core of ecosystem community models. Our core equations for a predator–prey system can be written in the form

$$\frac{dX_1}{dt} = r_1X_1 - \beta_{12}X_1X_2 \tag{14}$$

and

$$\frac{dX_2}{dt} = \beta_{21}X_1X_2 - r_2X_2. \tag{15}$$

If we consider X_1 the prey and X_2 the predator, then we can assume that the r_i and the βij are all positive constants.

The steady-state solutions to these equations are (a) $X_{2,0} = r_1/\beta_{12}$ and $X_{1,0} = r_2/\beta_{21}$ and (b) $X_{1,0} = X_{2,0} = 0$. The community matrix is now a 2×2 matrix, with these elements:

$$\mathbf{A}_{11} = \left.\frac{d[dX_1/dt]}{dX_1}\right|_{\text{steady state}} = r_1 - \beta_{12}X_2|_{\text{steady state}} \tag{16}$$

$$\mathbf{A}_{12} = \left.\frac{d[dX_1/dt]}{dX_2}\right|_{\text{steady state}} = -\beta_{12}X_1|_{\text{steady state}} \tag{17}$$

$$\mathbf{A}_{21} = \left.\frac{d[dX_2/dt]}{dX_1}\right|_{\text{steady state}} = \beta_{21}X_2|_{\text{steady state}} \tag{18}$$

$$\mathbf{A}_{22} = \left.\frac{d[dX_2/dt]}{dX_2}\right|_{\text{steady state}} = \beta_{21}X_1 - r_2|_{\text{steady state}}. \tag{19}$$

Consider, first, the steady state with nonzero population sizes. The community matrix is

$$\mathbf{A} = \begin{bmatrix} 0 & -\dfrac{r_2\beta_{12}}{\beta_{21}} \\ \dfrac{r_1\beta_{21}}{\beta_{12}} & 0 \end{bmatrix}. \tag{20}$$

The eigenvalues are given by the solution to the equation: determinant $(\mathbf{A} - \lambda\mathbf{I}) = 0$, or

$$\text{determinant} \begin{bmatrix} -\lambda & -\dfrac{r_2\beta_{12}}{\beta_{21}} \\ \dfrac{r_1\beta_{21}}{\beta_{12}} & -\lambda \end{bmatrix} = 0. \tag{21}$$

The determinant is readily evaluated, giving a quadratic equation for the eigenvalues:

$$\lambda^2 = -\frac{r_1\beta_{21}}{\beta_{12}}\frac{r_2\beta_{12}}{\beta_{21}} = -r_1r_2. \tag{22}$$

The square root of a negative number is imaginary, and so the two eigenvalues are

$$\lambda_1 = +i(r_1r_2)^{1/2} \tag{23}$$

and

$$\lambda_2 = -i(r_1r_2)^{1/2}. \tag{24}$$

This means that our steady state is not asymptotically stable—if the system is perturbed from the steady state, the state variables X_1 and X_2 will not return to their original value. Because the eigenvalues are imaginary, these perturbed solutions will, in fact, oscillate around the original steady state. This result is consistent with the results of the numerical analysis we carried out in Problem IV-4.

If the signs of β_{12} and r_1 are flipped, then our core model describes a pair of mutualists rather than a predator–prey pair. The interaction term augments the growth of each. In that case, the right side of the eigenvalue equation $\lambda^2 = -r_1r_2$ is a positive number, and the two solutions have the form $\lambda_1 = +(\text{positive number})^{1/2}$ and $\lambda_2 = -(\text{positive number})^{1/2}$. Now, because one eigenvalue has a positive real part, we have a genuine instability. It is easy to show that if the two species interact competitively, the same result holds.

The other steady state, the extinction state, gives the community matrix

$$\mathbf{A} = \begin{bmatrix} r_1 & 0 \\ 0 & r_2 \end{bmatrix} \tag{25}$$

and eigenvalues $\lambda_1 = r_1$ and $\lambda_2 = r_2$. Thus, the extinction state is unstable, just as it was for the logistic and Malthus models.

We argued in Problem IV-4 that introduction of density-dependent death rates into the core model was a step toward making the model more realistic. What does that density dependence do to the stability of the system? It can be shown that if a density-dependent death rate is inserted into either or both of Eqs. 14 and 15, then both eigenvalues have a negative real part. In other words, any amount of a density-dependent death rate stabilizes the core predator–prey model.

Interestingly, the interaction of mutualists or competitors is not so easily stabilized. In Exercise 3, you will get a chance to explore the stability of mutualism and competition models with density-dependent death rates.

The core of biogeochemical models. The equations for our core nitrogen model are

$$\frac{dX_1}{dt} = rX_1X_2 - fX_1{}^2 \tag{26}$$

$$\frac{dX_2}{dt} = bX_3 - rX_1X_2 \tag{27}$$

$$\frac{dX_3}{dt} = fX_1{}^2 - bX_3, \tag{28}$$

where $X_1 = P$, $X_2 = I$, and $X_3 = S_N$. Now we are dealing with a system of equations with a conservation law: $d[X_1 + X_2 + X_3]/dt = 0$, and as we shall see, this law affects the outcome of our stability analysis.

The steady states obey the relations

$$X_{2,0} = \frac{fX_{1,0}}{r} \tag{29}$$

and

$$X_{3,0} = \frac{X_{1,0}{}^2 f}{b}. \tag{30}$$

Because the three equations (26–28) are not all independent (the conservation of nitrogen links them), the three steady-state values, $X_{i,0}$, cannot be uniquely determined from the constants r, f, and b. If, however, the total quantity of nitrogen in the system, T_0, is specified, then in terms of T_0, the $X_{i,0}$ values can be determined. First, the equation $X_{1,0}(1+f/r)+X_{1,0}{}^2 f/b = T_0$ determines the value of $X_{1,0}$, and then the other two $X_{i,0}$ values are determined from Eqs. 29 and 30.

The community matrix for the model is

$$\mathbf{A} = \begin{bmatrix} rX_{2,0} - 2fX_{1,0} & rX_{1,0} & 0 \\ -rX_{2,0} & -rX_{1,0} & b \\ 2fX_{1,0} & 0 & -b \end{bmatrix}. \tag{31}$$

This matrix can be simplified somewhat using Eqs. 29 and 30:

$$\mathbf{A} = \begin{bmatrix} -fX_{1,0} & rX_{1,0} & 0 \\ -fX_{1,0} & -rX_{1,0} & b \\ 2fX_{1,0} & 0 & -b \end{bmatrix}. \tag{32}$$

The determinental equation for the eigenvalues reads

$$\text{determinant} \begin{bmatrix} -fX_{1,0} - \lambda & rX_{1,0} & 0 \\ -fX_{1,0} & -rX_{1,0} - \lambda & b \\ 2fX_{1,0} & 0 & -b - \lambda \end{bmatrix} = 0. \tag{33}$$

The determinant is readily evaluated, yielding

$$\begin{aligned} &(-fX_{1,0} - \lambda)(-rX_{1,0} - \lambda)(-b - \lambda) \\ &\quad + (2fX_{1,0})(rX_{1,0})(b) - (rX_{1,0})(-fX_{1,0})(-b - \lambda) = 0. \end{aligned} \tag{34}$$

This expression can be simplified to read

$$\lambda^3 + \lambda^2[(f+r)X_{1,0} + b] + \lambda[b(r+f)X_{1,0} + 2rfX_{1,0}{}^2] = 0. \tag{35}$$

One of the solutions to this equation is just

$$\lambda_1 = 0, \tag{36}$$

leaving us with a quadratic equation for the other two roots:

$$\lambda^2 + \lambda[(f+r)X_{1,0} + b] + [b(r+f)X_{1,0} + 2rfX_{1,0}{}^2] = 0. \tag{37}$$

The solutions to this equation are

$$\lambda_2 = \tfrac{1}{2}\{-[(f+r)X_{1,0} + b] + [(b-(f+r)X_{1,0})^2 - 8rfX_{1,0}{}^2]^{1/2}\} \tag{38}$$
$$\lambda_3 = \tfrac{1}{2}\{-[(f+r)X_{1,0} + b] - [(b-(f+r)X_{1,0})^2 - 8rfX_{1,0}{}^2]^{1/2}\}. \tag{39}$$

Because r, f, and b are positive quantities, we can easily show that the real parts of these two eigenvalues are negative. These eigenvalues will be real numbers if and only if $[b-(f+r)X_{1,0}]^2 \geq 8rfX_{1,0}{}^2$.

Before inserting typical numerical values for the parameters, we should see what general conclusions we can reach. Two of the eigenvalues must have negative real parts, so the terms in Eq. 13 in the Background section to this chapter that involve these two eigenvalues will damp out. The system would be asymptotically stable were it not for the first eigenvalue, which is zero (Eq. 36).

As I discussed in the Background, a zero eigenvalue implies neutral stability. This means that if the system is displaced from a steady state, it will not return to that state. Because the real part of λ_1 is zero and not positive, the deviation from the steady state will not expand exponentially over time, nor will it settle back to the unperturbed state. Upon an initial perturbation characterized by an addition of even a small amount of the conserved quantity to the system, the state variables can never return to their preperturbed values. This is a direct consequence of our conservation law. In particular, the existence of a zero eigenvalue is mathematically inevitable in a model in which the conservation law

$$\sum_i \frac{dX_i}{dt} \equiv 0 \tag{40}$$

holds. The reason is that Eq. 40 implies that the determinant of the community matrix vanishes, and thus the constant term (the term multiplying λ^0) vanishes. The variable λ can then be factored from every term in the determinant equation, and Eq. 36 follows.

To determine the actual response over time of the system variables, we have to look at the eigenvectors as well as the eigenvalues. The eigenvector corresponding to a particular eigenvalue, λ, is a linear combination of the variables, $\sum_i a_i X_i$, where the a_i are constants. It is the change over time of $\sum_i a_i \Delta X_i$ that has time dependence $\exp(\lambda t)$. Thus, it is of interest to see what the eigenvector looks like that corresponds to the eigenvalue $\lambda_1 = 0$. To answer this question, we turn to the eigenvector equation,

$$\mathbf{A}\mathbf{V}_\sigma = \lambda_\sigma \mathbf{V}_\sigma. \tag{41}$$

Because the eigenvalue with $\sigma = 1$ vanishes, this equation becomes

$$\mathbf{A}\mathbf{V}_\sigma = 0. \tag{42}$$

Hence, we have to solve the equation

$$\begin{bmatrix} -fX_{1,0} & rX_{1,0} & 0 & V_{1,1} \\ -fX_{1,0} & -rX_{1,0} & b & V_{1,2} \\ 2fX_{1,0} & 0 & -b & V_{1,3} \end{bmatrix} = 0. \tag{43}$$

This equation yields the following independent equations,

$$\frac{V_{1,2}}{V_{1,1}} = \frac{f}{r} \tag{44}$$

and

$$\frac{V_{1,3}}{V_{1,1}} = \frac{2f}{b}. \tag{45}$$

A third relationship can be extracted from Eq. 43, but it is equivalent to Eqs. 44 and 45.

Eqs. 44 and 45 tell us that the eigenvector, $\mathbf{V}_1$, is, up to an overall undeterminable constant, c,

$$\mathbf{V}_1 = \begin{bmatrix} c \\ \frac{cf}{r} \\ \frac{c2f}{b} \end{bmatrix}. \tag{46}$$

Now we can use Eq. 13 in the Background to this Chapter to see how a zero eigenvalue affects the response of the system variables to an initial displacement. If we let $t \to \infty$, the only surviving term in Eq. 13 is the one with $\sigma = 1$, corresponding to the zero eigenvalue. The other two terms in the sum over σ will vanish because $\exp(\lambda_2 t)$ and $\exp(\lambda_3 t)$ damp out at large values of t because the real parts of those eigenvalues are negative.

Hence, setting $\sigma = 1$ in Eq. 13 of the Background to Chapter V, we have

$$\begin{aligned}
\Delta X_1(t \to \infty) &\to C_{11}[(C^{-1})_{11}\Delta X_{2,0} + (C^{-1})_{12}\Delta X_{2,0} + (C^{-1})_{13}\Delta X_{3,0}] \\
\Delta X_2(t \to \infty) &\to C_{21}[(C^{-1})_{11}\Delta X_{2,0} + (C^{-1})_{12}\Delta X_{2,0} + (C^{-1})_{13}\Delta X_{3,0}] \\
\Delta X_3(t \to \infty) &\to C_{31}[(C^{-1})_{11}\Delta X_{2,0} + (C^{-1})_{12}\Delta X_{2,0} + (C^{-1})_{13}\Delta X_{3,0}].
\end{aligned} \tag{47}$$

These three terms all contain the exact same quantity in square brackets, and so the ratios of the three ΔX_i are just the ratios of the C_{i1} matrix elements. The matrix $\mathbf{C}$, as you recall, is composed of columns that are the eigenvectors of the community matrix. Hence, from Eq. 46, the vector of displacements from the steady state will obey the asymptotic relationships

$$\begin{aligned}
\frac{\Delta X_2}{\Delta X_1} &\to \frac{f}{r} \\
\frac{\Delta X_3}{\Delta X_1} &\to \frac{2f}{b}.
\end{aligned} \tag{48}$$

The eigenvector corresponding to the zero eigenvalue thus tells us the ratios of the state variables after the system has settled down to a new steady state long after an initial displacement. You should note the similarity between our result in Eq. 48, and in Eq. 11 of Problem IV-5, where some of the same information was obtained by a less formal method.

EXERCISE 1: Here you revisit Exercise 5 in Problem IV-2, the spread-sheet solution to the logistic equation, but you take larger values for the parameter r. In that exercise, you used $r = 0.01$, $K = 2$, and a unit time step. Now examine the cases $r = 1, 2, 2.2, 2.4, 2.6, 2.8$, and 3.0, all with $K = 2$ and unit time step. By numerically perturbing very slightly the initial state $X(0)$, you can examine the stability of this discrete-time version of the logistic equation model. What do you conclude by doing this? Is the stability of the differential equation form of the logistic model identical to that of the discrete form? The solutions you generated for the largest values of r exhibit a property called "chaos." Note that the solutions for these r values exhibit unpredictable behavior and are highly sensitive to initial conditions—those are the hallmarks of chaos. To learn more about this exciting phenomenon, consult "Population biology and ecology" under "Further Reading."

EXERCISE 2: There is a tendency for a population to slide to permanent extinction when the population size declines beyond a certain point. This slide can occur for a variety of reasons, including the difficulty animals face in finding mates at low population density (sometimes called the Allee Effect), the harm that can result from genetic inbreeding, and the increased likelihood that a freak weather pattern or some other stochastic event will wipe out all the individuals in a small population. Find a way to modify the logistic equation (see Problem IV-2) so that if the population declines below a critical value, then it slides toward the stable $X = 0$. At the same time, make sure that the model contains the stable state $X = K$.

EXERCISE 3: Consider the following pair of equations for two interacting species: $dX_1/dt = \alpha_1 X_1 - \gamma_1 X_1{}^2 + \beta_{12} X_1 X_2$ and $dX_2/dt = \alpha_2 X_2 - \gamma_2 X_2{}^2 + \beta_{21} X_1 X_2$, where the γ values, which set the strength of the density dependent death rates, are always positive. If the β values are both positive, these equations describe a pair of mutualists. If the β values are both negative and the α values are both positive, then they describe a pair of competitors. If, say, $\beta_{12} > 0$, $\beta_{21} < 0$, $\alpha_1 < 0$, and $\alpha_2 > 0$, then the equations describe a predator–prey pair that differs from the predator–prey pair we discussed above because of the presence of density dependence. For each of these cases, examine the stability of the system. For the first two cases, mutualism and competition, you will find that whether the real parts of the eigenvalues are negative or positive depends on the magnitudes of some of the parameters; try to characterize the conditions that make competition or mutualism stable.

EXERCISE 4: Suppose that the initial perturbation to our core biogeochemical model satisfies the condition that $\sum_i \Delta X_{i,0} = 0$. In other words, the conserved substance is neither added to, nor subtracted from, the system; instead it is rearranged among the system components. Determine the state of the system at $t = \infty$.

EXERCISE 5: Examine the stability of our Model 2 in Problem IV-5, but with a fixed value for S_C (so that you have a 4×4, not a 5×5, community matrix). Is the model always stable? Because of the conservation law, you should find that at least one of the eigenvalues is zero, leaving you with a cubic (rather than 4th order) polynomial equation for the eigenvalues. [*Hints*: You can determine if the roots of a polynomial all have negative real parts by applying the Routh–Hurwitz criteria, which work as follows. Let the polynomial equation for the eigenvalues be of the general form $a_n\lambda^n + a_{n-1}\lambda^{n-1} + \cdots + a_1\lambda + a_0 = 0$. The roots of this equation all have negative real parts if and only if certain constraints hold among the a_i. In particular, for the case $n = 3$, the Routh–Hurwitz necessary and sufficient conditions for stability are $a_0, a_3 > 0$ and $a_1a_2 > a_0$.]

EXERCISE 6: Consider a population of microorganisms living in organically poor lake water, and assume the population density obeys the logistic equation, $dN/dt = rN(1 - N/K)$, where N is the number of microorganisms per liter of water. Assume that $r = 1/\text{day}$ and $K = 10^8$. At time $t = 0$, the population is perturbed from its steady-state value ($N = K$) by the addition of organic matter. The effect of this addition is to suddenly and permanently raise the value of the carrying capacity, K, by 20%. Perturbations in a model parameter, such as K (and in contrast to perturbations in state variables X_i) are called structural perturbations. To analyze stability under a structural perturbation, we augment the model conveniently by pretending the perturbed parameter is a new state variable. Thus, a new equation, $d(\text{parameter})/dt = 0$, is added to the model; this equation expresses the constancy of the parameter, yet treats it as a state variable. Then a structural perturbation to the parameter can be treated as an ordinary perturbation to the new state variable, and eigenvalue analysis of the augmented community matrix can be used. The value of $\Delta X(0)$ in our equation for the solution to the community matrix problem (Eq. 13 in the Background to Chapter V) will be given by the change in K as given above in the problem statement. Using the matrix method for solving coupled differential equations, solve for the time-dependence of N subsequent to the addition of organic matter at $t = 0$.

V-2. Some Say in Ice

Ice and snow reflect a large fraction of incident sunlight and thus result in a cooler climate than would be the case if more sunlight were absorbed. This situation suggests the *possibility* of an instability, or runaway positive feedback, in which a small increase (or a small decrease) in the ice-covered area of the Earth cools (or warms) the climate and leads to the further increase (or decrease) of ice cover. Evaluate the feedback factor g in Eqs. 19 and 21 in the Background to Chapter V, and determine whether the Earth's ice caps are stable.

. . . .

We start with our simplest core climate model, Eq. 4 in Problem IV-3. This equation expresses the radiation balance of an atmosphere-free planet in the form

$$\frac{K\,dT}{dt} = \frac{\Omega}{4}(1 - a) - \sigma T^4. \tag{1}$$

Here, T is the average surface temperature of the planet, expressed in kelvin, K is the heat capacity per unit area of the planet, $\Omega/4$ is the incident solar flux averaged over the planetary surface, a is the albedo of the planet (i.e., the fraction of sunlight that is reflected back to space rather than absorbed and converted to heat), and σ is the Stefan–Boltzmann constant. Using a globally averaged value of $\Omega/4 = 343$ watts/m^2 and an initial value for albedo of $a = 0.30$ yields a value of $T \sim 255$ kelvin for the unperturbed globally averaged temperature.[31]

To tackle this problem, we have to partition the total albedo, a, into its contribution from ice or snow, and its contribution from all other sources. Because the feedback effect described in the problem statement involves a change in the area of the Earth that is covered by ice, let's introduce a variable, A, that is the fraction of Earth's surface area covered by ice. Currently, $A \cong 0.1$. In terms of A, the total albedo can be written

$$a = 0.7A + 0.1(1 - A) + 0.15. \tag{2}$$

31. Eq. 1 greatly simplifies the climate system, primarily because it ignores atmospheric effects; in Exercise 4, you will investigate stability against ice melt in a more realistic climate model in which radiation transfer within the atmosphere is included and that yields a more realistic, unperturbed value of $T \sim 290$ kelvin.

Here, 0.7 is the reflection coefficient of ice, and 0.1 is the average reflection coefficient of the non–ice-covered surface. The term 0.15 in this expression is the contribution to total albedo from the atmsophere (mostly clouds), and we assume this term is not affected by a change in A.[32]

The feedback arises because A is a function of temperature: if T increases, the fraction A shrinks, and this reduces the value of a. That, in turn, results in a larger value of T; and so it goes. To apply Eq. 21 in the Background to Chapter V, we express the steady-state solution to Eq. 1 in the form

$$T = G[a(T)] = \left\{\frac{\Omega}{4\sigma}[1 - a(T)]\right\}^{1/4}. \tag{3}$$

Using Eq. 2 (above), Eq. 3 can be rewritten

$$G[a(T)] = \left(\frac{\Omega}{4\sigma}\{1 - 0.7A(T) - 0.1[1 - A(T)] - 0.15\}\right)^{1/4}. \tag{4}$$

The feedback factor, g, is given by

$$g = \frac{\partial G}{\partial A}\frac{\partial A}{\partial T}. \tag{5}$$

Calculating the value of $\partial G/\partial A$ is straightforward—all we have to do is take a partial derivative of the right side of Eq. 4. To estimate the value of $\partial A/\partial T$, however, we revert to unabashedly crude guesswork. We estimate $\Delta A/\Delta T$ in the north polar region by comparing the coldest temperatures with those at the boundary between the present ice-covered and ice-free regions. This boundary occurs roughly at 60° latitude in the north, where the average temperature is approximately −3°C. At the North Pole, the temperature averages approximately −21°C, and hence for the *entire* northern polar ice mass to melt, we might assume as a simple and very rough estimate that the temperature would have to rise by about 18°C. A similar estimate ought to apply to sea ice in the south, but not necessarily to land-based ice on the Antarctic continent itself. The area of sea ice on Earth is somewhat more than twice as big as the area of the land-based ice on the

---

32. The terms in Eq. 2 add up to $0.7(0.1) + 0.1(0.9) + 0.15 = 0.31$, which is slightly greater than the value 0.30 usually cited for a. In Exercise 2, you will discover that the slight discrepancy is only partly the result of roundoff error; it is also a consequence of our neglect of a phenomenon called "multiple scattering." Moreover, the equation implicitly assumes that ice has the same albedo, regardless of latitude. That is not really true, however. The higher the latitude, the more oblique is the angle at which the sun's rays strike the ice, leading to higher albedo at higher latitude. The same is also true, of course, of the albedo of the nonice surface material that is exposed when ice melts.

Antarctic continent, so let's just ignore the latter and take

$$\frac{\Delta A}{\Delta T} \sim -\frac{0.1}{18} = -0.0055. \tag{6}$$

Note that we can also express this as -0.0055/kelvin because one kelvin is the same size unit as one degree Celsius.

Here is another, independent way to estimate the value of $\Delta A/\Delta T$. Unlike the method above, which was based on present-day spatial patterns in ice cover and temperature, this approach is based on temporal change as the ice melted at the end of the last ice age. The total ice-covered area of the Earth, including both continental ice and sea ice, is about 50×10^6 km^2 today, of which approximately 35×10^6 km^2 lies in the Northern Hemisphere. About 20,000 years ago, when the globally averaged surface temperature of the Earth was about 10°C colder, ice covered an *additional* area of $\sim 20 \times 10^6$ km^2 in the Northern Hemisphere. Thus, a total of $\sim 55 \times 10^6$ km^2 of ice covered the Northern Hemisphere. Our Eq. 6 predicts that a 10°C cooling from today should increase the Northern Hemisphere ice area by 55%, or from 35×10^6 km^2 to $\sim 54 \times 10^6$ km^2, in good agreement with data.

So, although our method of estimation is crude, it is probably in the right "ball park." As we shall see, the conclusions we reach about the stability of Earth's ice cover are sufficiently robust that our estimate of $\Delta A/\Delta T$ would have to be off by a factor of 3 for our general conclusion to be invalidated.

We now need to calculate the value of $\partial G/\partial A$. From Eq. 4,

$$\frac{\partial G}{\partial A} = \frac{1}{4} G(A)^{-3/4} \frac{\Omega}{4\sigma}(-0.6). \tag{7}$$

Using Eq. 3, Eq. 7 can be rewritten as

$$\frac{\partial G}{\partial A} = \frac{1}{4} T^{-3} \frac{T^4}{1-a}(-0.6) = -\frac{0.15T}{1-a}. \tag{8}$$

Combining Eqs. 6 and 8, we have

$$g = [-0.0055]\frac{-0.15T}{1-a} = +\frac{0.000825T}{1-a}. \tag{9}$$

This expression must be evaluated at the unperturbed temperature and albedo. Using a globally averaged value for unperturbed surface temperature of 255 kelvin (the unperturbed steady state that our simplified climate model yields), $A = 0.1$, and $a = 0.30$ gives us

$$g = 0.30. \tag{10}$$

As expected, $g > 0$, indicating positive feedback. From Eq. 19 in the Background to Chapter V, we can write

$$\text{Total Effect} = \frac{\text{Direct Effect}}{1-g} = 1.42 \times \text{Direct Effect}. \tag{11}$$

Hence, this feedback exerts an approximately 42% effect globally on the average temperature rise as a result of other driving forces.

Now let's look at the stability of the system by evaluating the eigenvalue of Eq. 1. Because albedo, a, depends on ice area, A (Eq. 2), and A depends on temperature, T, we have to make explicit the complete T-dependence of the right side of Eq. 1. Before substituting in a numerical relationship between A and T, it is useful for us to examine the general case of a linear dependence of A on T:

$$A = -\alpha T + \beta. \tag{12}$$

We have taken the constant α to be positive here and inserted the negative sign, knowing that ice area is a decreasing function of temperature. Substituting Eq. 12 into Eq. 2 and Eq. 2 into Eq. 1, we get

$$\begin{aligned}\frac{dT}{dt} &= \frac{1}{K}\left[\frac{\Omega}{4}(1 + 0.1 + 0.6\alpha T - 0.6\beta + 0.15) - \sigma T^4\right] \\ &= \frac{1}{K}\left[\frac{\Omega}{4}(1.25 - 0.6\beta + 0.6\alpha T) - \sigma T^4\right].\end{aligned} \tag{13}$$

The eigenvalue is determined from the derivative of this expression with respect to T, evaluated at the unperturbed steady-state solution:

$$\lambda = \left.\frac{d(dT/dt)}{dT}\right|_{T=[(\Omega/4\sigma)(1-a)]^{1/4}=255\ \text{kelvin}} = \left.\frac{1}{K}\left[0.6\alpha\frac{\Omega}{4} - 4\sigma T^3\right]\right|_{T=255}. \tag{14}$$

This eigenvalue is real and negative, indicating instability, if and only if

$$\begin{aligned}\alpha > \frac{4\sigma T^3}{0.6\Omega/4} &= \frac{6.67\sigma T^3}{\Omega/4} = 6.67\frac{1}{T}\frac{\sigma T^4}{\Omega/4} \\ &= \frac{6.7(1-a)}{T} = 0.018.\end{aligned} \tag{15}$$

How does the actual value of α compare with this critical value, above which the climate system is unstable against the ice–albedo feedback effect? From Eq. 6, $\alpha \sim 0.0055$, which is about 3 times smaller than the threshold value. Thus, this model predicts a healthy margin of stability under the positive feedback mechanism linking ice cover and temperature. Positive feedback does not necessarily imply instability.

From Eq. 13 in the Background to this chapter, we see that for a stable system, the eigenvalues determine the time dependence of recovery of the state variables when they are initially perturbed. For a system with only one state variable, such as our climate model, the interpretation of the eigenvalue is particularly simple: λ^{-1} is the exponential decay time for recovery. Consider, now, the value of λ^{-1} in the absence of the ice–albedo feedback mechanism, and then in its presence.

In the absence of feedback, $\alpha = 0$, and from Eq. 13 the eigenvalue (we will call it λ_0 because it is the value in the absence of the ice–albedo feedback mechanism) is given by

$$\lambda_0 = \frac{1}{K}[-4\sigma T^3]|_{T=255}. \tag{16}$$

In the presence of the feedback,

$$\lambda = \frac{1}{K}\left[0.6\alpha\frac{\Omega}{4} - 4\sigma T^3\right]\Bigg|_{T=255}. \tag{17}$$

The ratio, $r = \lambda/\lambda_0$, is

$$r = 1 - 0.6\alpha\frac{\Omega/4}{4\sigma T^3}. \tag{18}$$

Using $\sigma T^4 = \Omega(1-a)/4$, this equation can be rewritten as

$$r = 1 - \frac{0.6\alpha T}{4(1-a)}. \tag{19}$$

Referring to Eq. 9, which was evaluated for the special case of $\alpha = -0.0055$, we see that, in general, $g = -0.6\alpha T/4(1-a)$. Hence,

$$r = 1 - g. \tag{20}$$

Using Eq. 19 from the Background to this chapter, we see that the amplification factor, f, is related to r by

$$r^{-1} = \frac{1}{1-g} = f. \tag{21}$$

The ratio r^{-1} can also be related to the ratio of the recovery time for the system in the presence of feedback (τ) to the recovery time in its absence (τ_0):

$$r^{-1} = \frac{\lambda^{-1}}{\lambda_0{}^{-1}} = \frac{\tau}{\tau_0}. \tag{22}$$

This result, that $1/(1-g) = \tau/\tau_0$, is quite generally true, although it was derived here for the particular case of the ice–albedo feedback process within our core model. It holds whether the feedback factor, g, is positive or negative. A system with positive (negative) feedback will take more (less) time to recover from perturbation than a system with no feedback.

EXERCISE 1: Some have asserted that if ice ever covered a much larger fraction of the Earth's surface than the current value of $A \sim 0.1$, and in fact if it covered even more than during the last ice age when $A \sim 0.15$, then an instability could arise in which a little extra cooling would cause the entire Earth to ice over. Within the framework of our model, is this assertion true?

EXERCISE 2: The assumption implicit in the problem statement that total albedo equals the sum of surface albedo plus cloud albedo is not actually correct. It is easy to show (e.g., *COW-1*, pp. 89–94) that if a layer of clouds, with reflection coefficient a_C, sits above the

The BRIGHT side of global warming.

surface of the Earth (which has a reflection coefficient a_S), then the combined albedo of the cloud–surface system is $a = a_C + T^2 a_S / (1 - a_C a_S)$, where T is the transmission coefficient of the cloud layer. Qualitatively, what effect does this complication have on stability under a feedback process involving a change in A where a_S is, as before, a linear sum of $0.7A + 0.1(1 - A)$; i.e., does it make the system more or less stable? Assume $a_C = 0.15$ as before and that the original ice area is such that $a = 0.3$ as before.

EXERCISE 3: Our analysis leading to Eq. 14 neglected the possibility that K might depend on temperature and that this dependence could alter the stability of the system. Show that whatever the dependence of K on T, it can only influence the magnitude—not the sign—of the eigenvalue. Discuss qualitatively the likely sign of the change in K under climate warming. Then make a simple model of the dependence of K on T, and use your model to estimate the quantitative effect of this dependence on the magnitude of the eigenvalue.

EXERCISE 4: Explore how putting an atmosphere into the model affects the stability of the Earth's ice cover. To do this, refer to *COW-1*, pp. 160–171, where a simple model of the climatic consequences of the infrared-radiation-absorbing capacity of the Earth's atmosphere is developed and applied.

EXERCISE 5: Locally, in the polar regions, the effect of this feedback process could be much greater than we predicted above. This is the major reason that general circulation models predict that increasing atmospheric carbon dioxide levels will result in a much larger than average rise in temperature in the high latitudes. But if that is the case, we may also be underestimating the magnitude of g by treating the problem in the globally averaged manner that we did. Construct a "three-box" model of the Earth's surface, in which one box represents the present ice-free region and the other two represent the ice-covered polar regions. Using the data in the Appendix (Table XIV.1) to *COW-1*, assign each box an average temperature and a solar input, and couple the regions using the data in the Appendix on the average convective fluxes of energy across latitudinal bands. Estimate the magnitude of the feedback effect in each box and a globally averaged value for the feedback.

V-3. Biting the Hand that Feeds Us (II)

In Problem II-5, we looked at a very simple model of the implications of the relationships among economic activity, ecological health, and human well-being. Using that model, we found the level of economic activity that maximized human well-being. Pursue this topic further, now, and explore the stability of economic–ecological systems by examining a sequence of increasingly realistic models of such systems that capture (a) the tendency of ecosystems, at least under some circumstances, to recover from disturbance, (b) an exponentially growing economy, (c) the existence of a time lag between the state of the environment or a unit of economic activity and the effect of that state or activity on the rate of change of the environment, and (d) the presence of a critical threshold, beyond which an ecosystem will not recover. What general insights can you reach from analysis of your models?

• • • • • • •

The simple model we explored in Problem II-5 was formulated in terms of algebraic relations among the system components rather than with differential equations because we were not interested there in change over time. Here, however, time is of the essence. The model used in II-5 linked human welfare (W), economic output (Y), economic activity (X), and the health of the environment (Z). It looked as follows:

$$W = aYZ \tag{1}$$

$$Y = bXZ \tag{2}$$

$$Z = Z_0 - cX. \tag{3}$$

Feedback in the form of environmental degradation leading to impairment of the quality or quantity of resource input to economic activity is entirely described by the effect (Eq. 3) of economic activity, X, on the environmental quality, Z, and then the effect of Z on economic output, Y (Eq. 2). Because the phenomena listed under (a), (c), and (d) in the problem statement all refer in some way to the dynamics of ecosystems or, more generally, the environment, we will modify the form of Eq. 3 to incorporate those phenomena into the model.

Eq. 3 is deficient in a way that could cause it to seriously under- or overestimate the magnitude of loss of ecosystem services as a result of economic acitivity. In particular, it fails to capture the tendency for ecosystems to either heal or deteriorate progressively after being subjected to disturbance. Indicators of ecosystem health, such as measures of water quality, may decline under disturbance but subsequently recover if the stress is removed. The decline of lakewater alkalinity (see *COW-1*, p. 149) under acid deposition is an example, because natural alkalinity-restoring mechanisms such as mineral weathering have the potential, at least, to restore alkalinity to acidified ecosystems once acid deposition ceases. On the other hand, progressive deterioration may occur once an ecosystem has been stressed beyond some critical level. For example, when populations of organisms dwindle to sufficiently low levels, a combination of factors may lead the population to go rapidly extinct (see Problem V-1, Exercise 2). These factors include the increased likelihood of genetic inbreeding, the difficulty of organisms finding mates in a sparse population, and the increased vulnerability of a small population to extinction from a random catastrophic event (drought, early frost, blight, storm, fire, etc.). If populations go extinct, then ecosystem services will be lost as well.

Mechanisms of healing or progressive deterioration can be readily included in models that are only slightly more complex than Eqs. 1 through 3. To describe a dynamic environment, we simply replace Eq. 3 with a time-differential equation. The logistic equation provides a reasonable way to include "healing." For example, the logistic equation,

$$\frac{dZ}{dt} = rZ(t)[Z_0 - cX(t) - Z(t)], \tag{4}$$

predicts the eventual return of Z to the pristine value Z_0 once economic activity is turned off ($X = 0$), but the equation results in ecosystem degeneration toward a depressed value of Z (given by $Z_0 - cX$) while economic activity occurs.

The logistic equation is often used in ecology to model the time-dependence of a population subject to a finite carrying capacity, which it approaches when displaced by a disturbance. In our use of the equation here, we are going beyond the traditional application of the logistic model to populations and asserting that environmental indicators such as water quality or rate of methanotrophic activity in soil can also be usefully described by a logistic equation.

Our model now consists of Eqs. 1, 2, and 4. Before we examine the consequences of time lags and catastrophic thresholds, we first determine how two different assumptions about the time-dependence of economic activity, X, affect human welfare in this revised model that includes healing.

The effect of "ecosystem healing". First, consider a *steady-state economy*:

$$X = X_0 \quad \text{for } t \geq 0,$$
$$X = 0 \quad \text{for } t < 0.$$

If, for a long time prior to $t = 0$, X has been equal to 0, then Eq. 4 informs us that Z will have settled into a stable steady state at $Z = Z_0$ (see Problem V-1). So $Z(0) = Z_0$ is our initial condition, and now we have to solve Eq. 4 with $X(t) = X_0$ for the period $t > 0$. The solution to the logistic Eq. 4, subject to the initial condition, is (see Exercise 3 of Problem IV-2):

$$Z(t) = \frac{(Z_0 - cX_0)e^{rt}}{e^{rt} - cX_0/Z_0}. \tag{5}$$

Assuming $Z_0 > cX_0$, so that Z remains positive for all $t > 0$, this function declines monotonically from Z_0 to $Z_0 - cX_0$. Substituting this expression for Z into Eqs. 1 and 2, we derive the time-dependence of human welfare, $W(t)$, for $t \geq 0$,

$$W(t) = abX_0Z^2(t) \quad (t \geq 0). \tag{6}$$

As $t \to \infty$, this expression approaches $abX_0(Z_0 - cX_0)^2$, exactly the same answer we derived in Problem II-5. For $t < 0$, $X = 0$, and thus from Eqs. 1 and 2, our model tells us that

$$W(t) = 0 \quad (t < 0). \tag{7}$$

So nothing really new is emerging thus far from this dynamic model, other than that it yields the time-dependence characterizing the initial abrupt increase in welfare at $t = 0$, when economic activity starts up, and the gradual decline in W as Z drops from Z_0 to $Z_0 - cX_0$ (see Figure V-3).

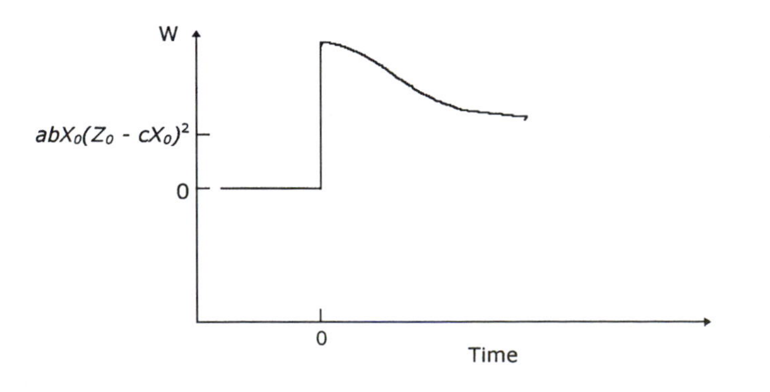

Figure V-3 Effect of an abrupt increase in economic activity on welfare.

Next, consider an *exponentially growing economy*:

$$X = X(0)e^{kt} \quad \text{for } t \geq 0,$$

$$X = 0 \quad \text{for } t < 0.$$

If economic activity, X, is exponentially growing for $t > 0$, then Eq. 4 is no longer analytically solvable for $Z(t)$. The general characteristics of the solution can be deduced readily from Eq. 4, however. As X grows exponentially, the term in brackets on the right side of Eq. 4 will shrink at an increasingly rapid rate and then become increasingly negative. This leads to an eventual rate of decline in Z that is slow at first and then rapidly increases. As Z approaches zero, however, dZ/dt also approaches zero, and thus the rate of decline slows and Z then asymptotically approaches $Z = 0$.

What we really care about is human welfare, $W(t) = abX(t)Z^2(t)$, and not economic output, X. Because $X(t)$ is an exponentially growing function of time and Z is a declining function of time, the behavior of XZ^2 over time is of interest. Figure V-4 shows a typical numerical output of the model for a particular set of values of the parameters: $a = b = 1$, $cX(0) = 0.1$, $r = 0.1$, and $k = 0.02$. Exercise 1 provides you the opportunity to explore how numerical choices for these parameters affect the model output.

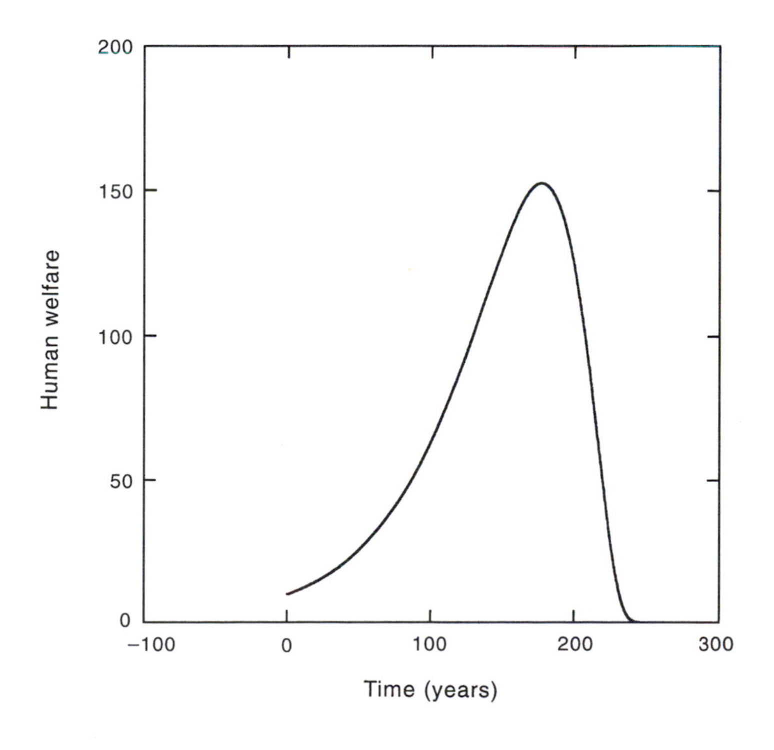

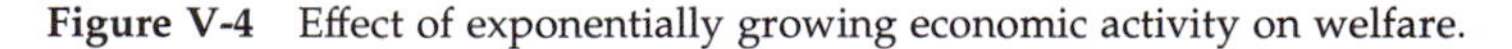

Figure V-4 Effect of exponentially growing economic activity on welfare.

The effect of time lags. In ecosystems, time lags can occur that cause a delayed response of a parameter (such as a population density or a chemical concentration) to the changes in the parameters that are coupled to it by interactions. Such lags can arise for various reasons, for example, (a) transport across space of toxic substances or nutrients or animals is not instantaneous; (b) eggs take time to hatch and organisms take time to grow from birth to reproductive age; and (c) a time lag exists between when a person becomes infected with a disease and when the person is capable of transmitting that disease to others.

To incorporate a time lag into our logistic model for $Z(t)$, we modify Eq. 4 as follows:

$$\frac{dZ}{dt} = rZ(t)[Z_0 - cX(t) - Z_L(t)], \tag{8}$$

where the subscript on Z_L stands for "lagged," and Z_L is given by

$$Z_L = \int_{-\infty}^{0} Z(t+\tau) f(\tau)\, d\tau. \tag{9}$$

Eqs. 8 and 9 represent just one of many ways (see Exercise 2) to express the rate of change of environmental quality as a function of the past state of the environment. The variable τ is called a "dummy variable." This means that it does not describe actual time in our model (i.e., the argument, t, of the variables Z, X, and W) but rather is a fictitious time variable we introduce so that we can describe the dependence of our variables on shifts in time; τ ranges from 0 (the present time in the model, which is time t) to $-\infty$ (the distant past). Thus, $Z(t+\tau)$ refers to Z at times shifted by an amount τ from the time t. The function $f(\tau)$ describes the influence of past environmental conditions on present environmental quality. It is normalized such that

$$\int_{-\infty}^{0} f(\tau)\, d\tau = 1. \tag{10}$$

For example, consider the particular lag function

$$f = \frac{1}{\tau_0} e^{\tau/\tau_0}. \tag{11}$$

This function falls off exponentially as the interval of time back into the past increases, has an average lag time of τ_0, and has its maximum at the present time ($\tau = 0$). Another example of a normalized lag function is

$$f = -\frac{\tau}{{\tau_0}^2} e^{\tau/\tau_0}. \tag{12}$$

This function is maximum at a time $t = -\tau_0$ (i.e., τ_0 time units before the present), and its average lag time is $2\tau_0$. It describes a situation in which the state of Z in the very recent past exerts less influence on the present rate of change of Z than does the state of Z at a time $-\tau_0$.

How can we analyze the stability of Eqs. 8 and 9? This time-lagged model does not appear in our taxonomy of equations in the Background to Chapter IV because it is not simply a differential equation. The reason is the presence of the integral in Eq. 9, which makes Eq. 8 an integrodifferential equation. In general, such equations are nasty beasts, often not easily solved or even studied by analytical methods. There is a trick, however, that permits us to study the stability of the equation, at least for a class of functions, $f(\tau)$. Assume a lag function in the form of Eq. 11. Then, taking the time derivative of Z_L, we get

$$\frac{dZ_L}{dt} = \int_{-\infty}^{0} \frac{d[Z(t+\tau)]}{dt} f(\tau)\, d\tau. \tag{13}$$

Because Z is a function of the linear combination $t + \tau$, the derivative d/dt can be replaced by the derivative $d/d\tau$ and the equation rewritten as

$$\frac{dZ_L}{dt} = \int_{-\infty}^{0} \frac{d[Z(t+\tau)]}{d\tau} f(\tau)\, d\tau. \tag{14}$$

Now we can integrate by parts, writing this equation as

$$\begin{aligned}\frac{dZ_L}{dt} = &-\int_{-\infty}^{0} Z(t+\tau)\frac{d[f(\tau)]}{d\tau}\, d\tau \\ &+ Z(t+\tau)f(\tau)|_{\tau=0} - Z(t+\tau)f(\tau)|_{\tau=-\infty}.\end{aligned} \tag{15}$$

Because f vanishes at $\tau = -\infty$, the third term on the right side vanishes; because $f(0) = 1/\tau_0$, the second term is given by $Z(t)/\tau_0$. Hence,

$$Z(t+\tau)f(\tau)|_{\tau=0} - Z(t+\tau)f(\tau)|_{\tau=-\infty} = \frac{Z(t)}{\tau_0}. \tag{16}$$

Now consider the first term on the right side of Eq. 15. From Eq. 11, the derivative in the integrand is given by $d[f(\tau)]/d\tau = (1/\tau_0)f(\tau)$. Inserting this term into the first term yields

$$\begin{aligned}\int_{-\infty}^{0} Z(t+\tau)\frac{f(\tau)}{d\tau}\, d\tau &= \frac{1}{\tau_0}\int_{-\infty}^{0} Z(t+\tau)f(\tau)\, d\tau \\ &= \frac{1}{\tau_0} Z_L.\end{aligned} \tag{17}$$

Hence, combining Eqs. 15, 16, and 17,

$$\frac{dZ_L}{dt} = -\frac{Z_L}{\tau_0} + \frac{Z(t)}{\tau_0}. \tag{18}$$

Collecting these results, we find that Eqs. 8 and 9, with f given by Eq. 11, is equivalent to the following family of equations:

$$\frac{dZ}{dt} = rZ(t)[Z_0 - cX(t) - Z_L(t)] \tag{19}$$

$$\frac{dZ_L}{dt} = -\frac{Z_L(t)}{\tau_0} + \frac{Z(t)}{\tau_0}. \tag{20}$$

Consider the case in which $X(t) = X_0$, a constant, with $Z_0 > X_0$. To simplify the notation, we let the positive parameter $b = r(Z_0 - X_0)$, so that Eq. 19 becomes

$$\frac{dZ}{dt} = bZ(t) - rZ(t)Z_L(t). \tag{19'}$$

The steady-state solutions to Eqs. 20 and 19′, which we denote by Z and $Z_{L,0}$ are $Z = Z_{L,0} = b/r$. We have seen previously (Eq. 7 of Problem IV-2) that the ordinary unlagged logistic equation, $dZ/dt = bZ - rZ^2$, has an eigenvalue given by $\lambda = -b$. Because this eigenvalue has a negative real part, the equation is stable against small perturbations. Moreover, because the eigenvalue has zero imaginary part, the perturbed solution recovers monotonically to its unperturbed state; if the eigenvalue had an imaginary part, the perturbed solution would oscillate about the steady state. How does time-lagging affect stability? To answer this question, we have to calculate the eigenvalues of the pair of Eqs. 19′ and 20. The community matrix of this pair of equations is:

$$\mathbf{A} = \begin{bmatrix} 0 & -b \\ \frac{1}{\tau_0} & -\frac{1}{\tau_0} \end{bmatrix}. \tag{21}$$

The eigenvalues of this matrix are given by the roots of the equation: determinant $(\mathbf{A} - \lambda\mathbf{I}) = 0$, which can be written (see the example of eigenvalue calculation in the Appendix)

$$\lambda^2 + \frac{\lambda}{\tau_0} + \frac{b}{\tau_0} = 0. \tag{22}$$

This equation has two solutions:

$$\lambda_1 = \frac{1}{2}\left[-\frac{1}{\tau_0} + \frac{1}{\tau_0}(1 - 4b\tau_0)^{1/2}\right] \tag{23}$$

and

$$\lambda_2 = \frac{1}{2}\left[-\frac{1}{\tau_0} - \frac{1}{\tau_0}(1 - 4b\tau_0)^{1/2}\right]. \tag{24}$$

We examine each of these eigenvalues in two limiting situations: τ_0 very small and τ_0 very large. The limit $\tau_0 \to 0$ corresponds to a vanishing lag, and so we ought to recover the original eigenvalue: $\lambda = -b$; let's see if we do. Using $(1-\varepsilon)^{1/2} \sim 1 - \varepsilon/2$ for $\varepsilon \ll 1$, we get for the limit $\tau_0 \to 0$:

$$\lambda_1 \sim \frac{1}{2}\left[-\frac{1}{\tau_0} + \frac{1}{\tau_0}\left(1 - \frac{4b\tau_0}{2}\right)\right] = -b + \text{terms of order } \tau_0 \text{ or smaller} \tag{25}$$

$$\lambda_2 \sim \frac{1}{2}\left[-\frac{1}{\tau_0} - \frac{1}{\tau_0}(1 - 4b\tau_0)^{1/2}\right] = -\frac{1}{\tau_0} \to -\infty. \tag{26}$$

Thus, we see that in this limit, we do indeed recover the time dependence of the unlagged model. In particular, the eigenvalue λ_1 corresponds to the eigenvalue for the unlagged model; because the other eigenvalue goes to $-\infty$, Eq. 13 in the Background tells us that it makes no contribution to the behavior of the variable $Z(t)$.

In the limit of large τ_0,

$$\lambda_1 = \frac{1}{2}\left[-\frac{1}{\tau_0} + \frac{1}{\tau_0}(1 - 4b\tau_0)^{1/2}\right] \to -\tau^{1/2}(-b)^{1/2} \to -i\infty \tag{27}$$

$$\lambda_2 = \frac{1}{2}\left[-\frac{1}{\tau_0} - \frac{1}{\tau_0}(1 - 4b\tau_0)^{1/2}\right] \to +\tau^{1/2}(-b)^{1/2} \to +i\infty \tag{28}$$

Hence, in this limit, the eigenvalues are both purely imaginary, and thus $Z(t)$ is no longer stable against an initial displacement—Eq. 14 in the Background tells us that the system undergoes neutrally stable oscillations after an initial displacement.

What happens if τ_0 is neither very small nor very large? The eigenvalues become complex (that is, the imaginary parts of the eigenvalues are nonzero, and thus the perturbed solutions oscillate) if $\tau_0 > (4b)^{-1}$. The table below presents a few values of the real part of the eigenvalues for particular values of $1/\tau_0$.

$1/\tau_0$	Re(λ_1)	Re(λ_2)
$32b$	$-16b[1-(7/8)^{1/2}] \cong -1.03b$	$-16b[1+(7/8)^{1/2}] \cong -31.0b$
$16b$	$-8b[1-(3/4)^{1/2}] \cong -1.07b$	$-8b[1+(3/4)^{1/2}] \cong -14.9b$
$8b$	$-4b[1-(1/2)^{1/2}] \cong -2.83b$	$-4b[1+(1/2)^{1/2}] \cong -6.83b$
$4b$	$-2b$	$-2b$
$2b$	$-b$	$-b$
b	$-b/2$	$-b/2$
$b/2$	$-b/4$	$-b/4$

We note that if $\tau_0 > (2b)^{-1}$, then the real part of the eigenvalues is less negative than $-b$, and thus the system recovers more slowly from a disturbance than in the unlagged case ($\tau_0 = 0$). Thus, in this situation, the time lag has rendered the system less stable. For very small τ_0, the real parts of the eigenvalues are more negative than $-b$, and thus the time it takes for a disturbance to dampen actually decreases. The critical transition point occurs at $\tau_0 = (2b)^{-1}$. The eigenvalues never develop positive real parts, however, and so a genuine instability never occurs.

The main lesson to be extracted from the above example is that time lags can reduce the stability of a system if the past is weighted strongly enough. The detailed effects of a time lag depend upon exactly where the time lag is inserted into the logistic equation and on the form of the lag function (Eq. 12 vs. Eq. 11, for example). If the lag function weights the past strongly enough, as Eq. 12 can do, then the time-lagged system can actually become unstable. Exercises 2 and 4 provide an opportunity to explore these issues further.

The effect of critical recovery thresholds. Finally, we turn to the last of the issues raised in the problem statement: the effect of a critical threshold beyond which an ecosystem will not recover. A variety of simple modifications of Eqs. 1, 2, and 4 allow us simultaneously to include progressive deterioration in the model. If the parameter, r, that characterizes the recovery rate in Eq. 4 is replaced by a function of Z that decreases sufficiently rapidly when Z is small, then recovery may be impossible when Z dips below a critical level. For example, the modified logistic equation,

$$\frac{dZ}{dt} = Z[r - s(Z - Z_0)^2](Z_0 - cX - Z) \tag{29}$$

with $r < sZ_0{}^2$, describes a system that can change rapidly from the state $Z = Z_0 - cX$ to the permanent state $Z = 0$ when cX exceeds a critical value. In other words, if the effective state of the environment, $Z_0 - cX$, is depressed too much because of too high a level of economic activity, the ecosystem collapses to $Z = 0$. This critical stability threshold value for economic activity, X_c, is given by (Exercise 5)

$$X_c = \frac{1}{c}\sqrt{\frac{r}{s}}. \tag{30}$$

To determine a numerical value for this stability threshold, we would need not only numerical values for the ecosystem parameters r and s (for which information is available for at least some ecosystems), but also the economic–ecological data to determine the constant c. Although we are currently unable to put forth a reliable numerical value for this threshold for any particular economic–environmental system, it would be worth developing the capability to make such an estimate and compare the resulting value of X_c with the current level of economic activity.

Eq. 29 also predicts that as X increases from 0 toward the critical stability threshold value given in Eq. 30, the sensitivity of the ecosystem to natural stochastic or deterministic stresses (such as interannual variations in precipitation, outbreaks of diseases, introduction of exotic organisms, etc.) increases dramatically. Thus, even before the level of economic activity is so high that the environmental system collapses, its vulnerability to perturbation increases. The measure of this sensitivity is the time constant characterizing the recovery of the system from such perturbations. The relevant time constant is given by (Exercise 6)

$$T_{\text{recovery}} = \frac{1}{r - s(cX)^2}. \tag{31}$$

Moreover, with increasing recovery time, the chances of a random perturbation pushing the system to the point of collapse (the state $Z = 0$) rapidly increases.

Finally, the model described by Eq. 29 can be further modified by replacing the term cX with the time-lagged term: $\int_{-\infty}^{0} X(t + \tau)f(\tau)\,d\tau$. Even if X is constant in time, the effect of this substitution is to reduce further the stability of the system; in particular, the recovery time following a disturbance is lengthened, and the onset of instability lies at a smaller value of X. Moreover, if X is time-dependent—for example, the periodic behavior caused by economic cycles—then the lagged logistic equation can lead to a variety of behaviors for Z, including chaotic solutions (Exercise 7).

EXERCISE 1: Set up Eqs. 1, 2, and 4, along with $X(t) = X(0)e^{kt}$ and $X = 0$ for $t < 0$, on a spread sheet and then fool around with the parameters a, b, $cX(0)$, k, and r to determine the range of behaviors of $W(t)$ that are possible in our exponentially growing economy model.

EXERCISE 2: Eq. 8 is only one of many ways to insert a time lag into the logistic equation. Using our integration-by-parts trick and the lag function given by Eq. 11, derive the effect on stability of lagging the model in the following way: $dZ/dt = rZ_L(Z_0 - cX - Z)$. What about $dZ/dt = rZ_L(Z_0 - cX - Z_L)$? Try to identify and describe a situation that each of these lagged models might describe.

EXERCISE 3: Consider the case of an exponentially growing $X(t)$, but with the effect of X on Z lagged by the function f in Eq. 11. In other words, let $dZ/dt = rZ(Z_0 - cX_L - Z)$, where $X_L = \int_{-\infty}^{0} X(t + \tau)f(\tau)\,d\tau$. Using this lagged model for Z, along with Eqs. 1 and 2, evaluate numerically with a spread sheet the time dependence of $W(t)$, and compare it with the time dependence of W given in Figure V-3.

EXERCISE 4: Determine the stability of the lagged logistic model (Eq. 8) using the lag function given by Eq. 12. [*Hint*: You will have to apply our integration-by-parts trick twice for this lag function; first introduce Z_L, which equals the lagged integral of Z over the f in Eq. 12, and then introduce a new variable, Z_{LL}, equal to a lagged integral of Z_L over the function f in Eq. 11. You will need to use the Routh–Hurwitz criteria for negative real parts of the roots of a cubic equation, as presented in Exercise 5 of Problem V-1.]

EXERCISE 5: Show that Eq. 30 follows from Eq. 29.

EXERCISE 6: Derive Eq. 31. [*Hint*: Recall from the Background to Chapter V the relationship between eigenvalues and recovery time constants.]

EXERCISE 7: Using a spread-sheet approach, examine the solutions to the lagged critical threshold model; over what ranges of numerical values for the key parameters is there evidence for steady states, periodicity, and chaos?

V-4. Sagebrush World

A model called "Daisy World" has been proposed[33] to bolster the case that the planetary ecosystem is adapted, like an organism, to be buffered against large-scale disturbances. The model describes dark and light flowers, each with a different temperature tolerance, and is rigged to result in negative feedback that stabilizes the climate–ecosystem complex. The feedback is a consequence of the fact that dark objects placed in sunlight warm up more than light objects.

If this book has convinced you of anything, perhaps you now see the flexible capability of models to describe many different situations. Our ability to create a model whose output looks like *X* has absolutely no bearing on whether the world looks like *X*. So let's see what the data suggest with regard to potential ecosystem–climate feedback—let's look at "Sagebrush World."

Experimental evidence suggests that under climate warming, meadow ecosystems may undergo shifts in the composition of their plant communities, in which plants like sagebrush (*Artemisia tridentata*) do better and plants like *Erigeron speciosus* (a daisy) do worse. Radiometric measurements indicate that different plant species do indeed have different albedos—darker plants absorb more solar energy than lighter ones (also, the albedo of a given species often varies with time of year and local meteorologic conditions). Sagebrush growing at higher elevations tends to have a lower albedo than does *Erigeron*, so its success creates warmer conditions that are more favorable for sagebrush and less favorable for *Erigeron*. Under what conditions might this positive feedback be unstable?

. . . .

33. Watson, A. J., Lovelock, J. E. 1983. Biological homeostasis of the global environment: the parable of daisyworld. *Tellus* 35B:284–89.

We will tackle this problem for the general case first, and then plug in reasonable numerical values for parameters. Like the ice–albedo feedback problem (V-2), this problem involves areas of differing albedos; unlike that problem, the areas of the surfaces of differing albedos can grow or shrink as a consequence of biological processes. Thus, this problem involves looking at feedback and stability in a model that combines our core climate and our core ecological models. Our knowledge of the interactions of the two plant species is not global but rather at the scale of a meadow, so we will have to think about how to adapt our climate model so that it is descriptive of the local, not necessarily the global, climate.

Let A_D and A_L be the areas covered by the darker (low albedo) and the lighter (high albedo) plant species, respectively, and let the total area of the meadow be A_0. Then, the total albedo of the meadow in which both species coexist can be written as

$$a = a_L \frac{A_L}{A_0} + a_D \frac{A_D}{A_0} + a_1 \frac{A_0 - A_L - A_D}{A_0}. \qquad (1)$$

Here a_L and a_D are the albedos of each of the two plant species, and a_1 refers to the albedo of all the rest of the meadow not covered by either of our two plant species.

Because our population models are generally based on changes in biomass, we let X_D and X_L refer to the biomasses of the two species in the meadow, but we assume that these biomasses are proportional to the occupied areas, so that

$$X_D = c_D A_D \qquad (2)$$

and

$$X_L = c_L A_L. \qquad (3)$$

Our climate model (from Problem IV-3) is

$$\frac{K\,dT}{dt} = \frac{\Omega}{4}(1 - a) - \sigma T^4. \qquad (4)$$

To this equation, we must couple a population model for the plants. First, we look at a plant model without competition, in which the two plant species independently respond to temperature but not to each other. This model consists of two uncoupled logistic equations with carrying capacities and/or growth rates dependent on temperature. Then we set up a Lotka–Volterra model that can describe the competition between the species.

Global warming sparks another sagebrush rebellion.

Model 1: No competition. We assume that each of the plant species obeys a logistic equation:

$$\frac{dX_D}{dt} = r_D X_D (K_D - X_D) \tag{5}$$

and

$$\frac{dX_L}{dt} = r_L X_L (K_L - X_L). \tag{6}$$

To capture in our model the observation that increasing temperature provides an advantage to the darker plants (the sagebrush) and hurts the lighter plants (the daisies), we set the r and K variables to be functions of temperature.[34] We adopt the simplest possible approach and take the parameters to be linear functions of temperature:

$$r_D = r_{D,0} + r_{D,1}T \tag{7}$$

$$r_L = r_{L,0} + r_{L,1}T \tag{8}$$

$$K_D = K_{D,0} + K_{D,1}T \tag{9}$$

$$K_L = K_{L,0} + K_{L,1}T. \tag{10}$$

By taking $r_{D,1}$ and $K_{D,1} > 0$ and $r_{L,1}$ and $K_{L,1} < 0$, we build into our model the temperature preferences of the two species.

34. At the end of the analysis, you will be able to see whether our results depend on there being temperature dependence in both the r and K variables, in just the r, or in just the K.

Our model thus consists of the three differential equations, Eqs. 4 through 6, supplemented by Eqs. 1 through 3 and 7 through 10. To simplify the notation, we let

$$q_{\mathrm{L}} = (Kc_{\mathrm{L}}A_0)^{-1}\frac{\Omega}{4}(a_{\mathrm{L}} + a_1) \tag{11}$$

$$q_{\mathrm{D}} = (Kc_{\mathrm{D}}A_0)^{-1}\frac{\Omega}{4}(a_{\mathrm{D}} + a_1) \tag{12}$$

$$p_0 = K^{-1}\frac{\Omega}{4}(1 - a_1) \tag{13}$$

and

$$w = K^{-1}\sigma. \tag{14}$$

In terms of these new parameters, Eq. 4, supplemented by Eqs. 1 through 3, can be written

$$\frac{dT}{dt} = p_0 - q_{\mathrm{L}}X_{\mathrm{L}} - q_{\mathrm{D}}X_{\mathrm{D}} - wT^4. \tag{15}$$

The community matrix for the three differential equations, Eqs. 5, 6, and 15, can be readily evaluated:

$$\mathbf{A} = \begin{bmatrix} -r_{\mathrm{D}}K_{\mathrm{D}} & 0 & K_{\mathrm{D},1}X_{\mathrm{D},0}r_{\mathrm{D}} \\ 0 & -r_{\mathrm{L}}K_{\mathrm{L}} & K_{\mathrm{L},1}X_{\mathrm{L},0}r_{\mathrm{L}} \\ -q_{\mathrm{D}} & -q_{\mathrm{L}} & -4wT_0{}^3 \end{bmatrix}, \tag{16}$$

where $X_{\mathrm{D},0}$, $X_{\mathrm{L},0}$, and T_0 are the steady-state values of the variables X_{D}, X_{L}, and T that result from setting the time derivatives in Eqs. 5, 6, and 11 (subject to the constraints of Eqs. 7–10) equal to zero. From this matrix, the following characteristic equation for the eigenvalues results:

$$(\lambda + 4wT_0{}^3)(\lambda + r_{\mathrm{D}}K_{\mathrm{D}})(\lambda + r_{\mathrm{L}}K_{\mathrm{L}}) + (\lambda + r_{\mathrm{D}}K_{\mathrm{D}})K_{\mathrm{L},1}X_{\mathrm{L},0}r_{\mathrm{L}}q_{\mathrm{L}} + (\lambda + r_{\mathrm{L}}K_{\mathrm{L}})K_{\mathrm{D},1}X_{\mathrm{D},0}r_{\mathrm{D}}q_{\mathrm{D}} = 0. \tag{17}$$

This equation can be rewritten as

$$\lambda^3 + (4wT_0{}^3 + r_{\mathrm{L}}K_{\mathrm{L}} + r_{\mathrm{D}}K_{\mathrm{D}})\lambda^2 + [4wT_{\mathrm{o}}{}^3(r_{\mathrm{L}}K_{\mathrm{L}} + r_{\mathrm{D}}K_{\mathrm{D}}) + r_{\mathrm{L}}K_{\mathrm{L}}r_{\mathrm{D}}K_{\mathrm{D}} + K_{\mathrm{L},1}X_{\mathrm{L},0}r_{\mathrm{L}}q_{\mathrm{L}} + K_{\mathrm{D},1}X_{\mathrm{D},0}r_{\mathrm{D}}q_{\mathrm{D}}]\lambda + r_{\mathrm{D}}r_{\mathrm{L}}(K_{\mathrm{L},1}X_{\mathrm{L},0}q_{\mathrm{L}} + K_{\mathrm{D},1}X_{\mathrm{D},0}q_{\mathrm{D}}) = 0. \tag{18}$$

By the Routh–Hurwitz criteria (see Exercise 5 of Problem V-1), a *sufficient* condition for at least one of the eigenvalues to have a positive real part, and thus for the system to be unstable, is that the term $r_D r_L(K_{L,1}X_{L,0}q_L + K_{D,1}X_{D,0}q_D) < 0$. (Exercise 3 gives you a chance to explore the implications of the other Routh–Hurwitz criteria, and thus to explore possibly weaker conditions for the instability of our system.) Because the r values are positive, this is equivalent to the condition that the quantity

$$\theta \equiv (K_{L,1}X_{L,0}q_L + K_{D,1}X_{D,0}q_D) < 0. \tag{19}$$

Because of the differing temperature responses in the growth of the two species, we know that $K_{L,1} < 0$ and $K_{D,1} > 0$. Hence, θ will be negative, and the system will be unstable, if $|K_{L,1}|/K_{D,1} > q_D X_{D,0}/q_L X_{L,0}$. To simplify matters, let's assume for now (but see Exercise 1) that all the land area is occupied by either sagebrush or daisies (i.e., $a_1 = 0$). Eqs. 11 and 12 then tell us that $q_D/q_L = c_L a_D/c_D a_L$, and so the instability condition is

$$\frac{|K_{L,1}|}{K_{D,1}} > \frac{a_D X_{D,0}/c_D}{a_L X_{L,0}/c_L}. \tag{20}$$

Using Eqs. 2 and 3, this equation can be rewritten as

$$\frac{|K_{L,1}|}{K_{D,1}} > \frac{a_D A_{D,0}}{a_L A_{L,0}}, \tag{21}$$

where $A_{D,0}$ and $A_{L,0}$ are the initial steady-state values of the areas occupied by the two types of plants. Because the sagebrush has a lower albedo than the daisies, a_D is less than a_L, and thus our instability condition is relatively weak—it does not require that $|K_{L,1}|$ be greater than $K_{D,1}$. To the extent that the area initially covered by sagebrush is less than that covered by daisies, this threshold condition for instability is even weaker. In other words, Sagebrush World can be highly vulnerable to instability!

Model 2: Competition. Competition between the sagebrush and the daisies can be included in our model by augmenting Eqs. 5 and 6 as follows:

$$\frac{dX_D}{dt} = r_D X_D(K_D - X_D - \beta_{DL} X_L) \tag{22}$$

and

$$\frac{dX_L}{dt} = r_L X_L(K_L - X_L - \beta_{LD} X_D). \tag{23}$$

When the two coefficients β_{LD} and β_{DL} are positive they are called "competition coefficients"; if they are negative, they describe a mutualism between the two species; if one is zero and the other is negative, the relationship is said to be "commensal." Even though plants do not eat each other and therefore are not linked by a predator–prey relationship, there are situations in which the two β values could have opposite signs, for example, if one species shades the other, thereby reducing available sunlight and stunting the other's growth, but the shaded one is a legume that provides nitrogen for the first.[35]

The stability of Eqs. 11, 22, and 23 can be analyzed just as we did for the no-competition model. This analysis is left as Exercise 4 for you.

EXERCISE 1: In deriving the sufficient condition for instability of the no-competition model (Eq. 21), we assumed that $a_1 = 0$. If $a_1 > 0$, would our instability condition be weaker or stronger?

EXERCISE 2: Substitute reasonable numbers for all the parameters in our no-competition model, and run simulations using a spread-sheet approach.

EXERCISE 3: Analyzing the characteristic Eq. 18 for the eigenvalues, we used only one of the Routh–Hurwitz conditions: $a_3 > 0$. What do the other conditions (see Exercise 5 of Problem V-1) tell us about the stability of our no-competition model? Could an even weaker condition than Eq. 20 lead to instability, assuming $c_D = c_L$?

EXERCISE 4: Examine the stability of the competition model. Does the presence of competition increase or decrease the likelihood that Sagebrush World is unstable?

EXERCISE 5: Take a set of numerical parameters that do not satisfy our instability condition but do correspond to positive feedback: i.e., $(K_{L,1}X_{L,0}q_L + K_{D,1}X_{D,0}q_D) > 0$; $K_{L,1} < 0$ and $K_{D,1} > 0$; $a_D < a_L$. Work out the feedback gain factor, $g = (1 - f)^{-1}$, and estimate how much additional temperature increase this positive feedback could cause if some external cause of warming, such as the "greenhouse effect," results in an initial warming of 1°C.

---

35. The interactions among plant species can be complex, indeed. Sagebrush is deep-rooted and can draw water up from greater soil depth, where water is more plentiful, and then release the water in the shallow root zone where other plants, or perhaps ammonifying bacteria, can use it.

V-5. Will the Seas Go Flat?

After sitting for an hour at room temperature, a glass of beer will be "flatter" than if it remained cool. Our oceans contain about 50 times as much carbon dioxide (including its dissociated ionic forms, bicarbonate and carbonate) as does the atmosphere, and thus loss of a sizeable fraction of oceanic carbon dioxide to the atmosphere could have an enormous effect on the atmospheric level of this greenhouse gas. Could the oceans go flat (i.e., lose a large amount of carbon dioxide to the atmosphere) because of the following positive feedback loop?

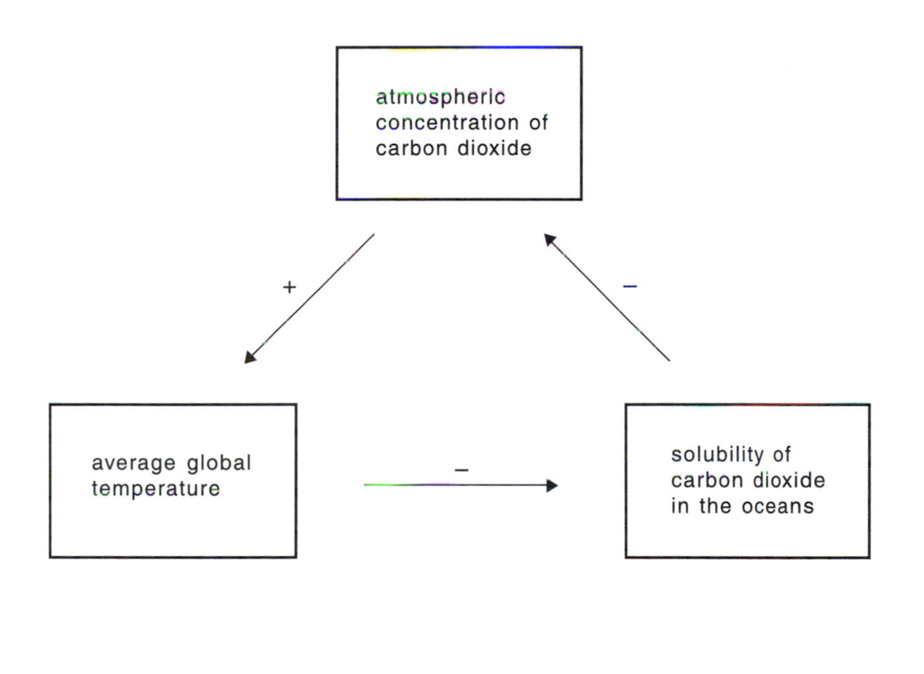

. . . .

The instability that such a positive feedback effect could induce has been called a "runaway greenhouse effect." If it occurred, our climate would be changed drastically by the release of the CO_2 in our oceans. Our task here is to see if earth could undergo such a catastrophe.

Let's first make sure we understand the mechanism at work here—why did the warm beer go flat faster? The equilibrium relationship between the concentration of dissolved carbon dioxide, $[H_2CO_3]$, and the partial pressure of atmospheric carbon dioxide, $p(CO_2)$, is described by a parameter called Henry's constant:[36]

$$[H_2CO_3] = K_{H,CO2}p(CO_2). \tag{1}$$

$[H_2CO_3]$ is measured in units of moles per liter, and $p(CO_2)$ is measured in atmospheres. Henry's constant for CO_2, $K_{H,CO2}$, is numerically about $10^{-1.47}$ moles/liter-atmosphere. In equilibrium with the atmosphere, cold water holds more carbon dioxide than warm water in part because Henry's constant for CO_2 is not really a constant, it is a decreasing function of temperature. This is the reason for the negative sign on the arrow in the feedback loop connecting average global temperature and the solubility of CO_2 in the oceans. The second negative sign, on the arrow connecting solubility of CO_2 in the oceans to the concentration of CO_2 in the atmosphere, is a consequence of mass balance—the carbon released from the oceans has to go somewhere, and the major sink for this carbon is the atmosphere.

Finally, there is a plus sign on the arrow connecting atmospheric CO_2 concentration to global temperature. This is just the greenhouse effect. As the atmospheric CO_2 level rises, the atmosphere absorbs and re-radiates more heat radiation; this effect implies that some additional heat radiation from Earth's surface, which would otherwise be radiated to space, is trapped and re-radiated back to Earth's surface, where it causes warming.

The overall sign of this feedback loop is positive $[(-)(-)(+) = +]$, and so we might have the potential for a runaway greenhouse effect. Concern over this possibility is heightened by the fact that the oceans could potentially release about 50 times more CO_2 than is found in the present atmosphere, so a huge buildup of atmospheric CO_2 could perhaps be triggered by this mechanism.

When carbon dioxide enters seawater, it will undergo inorganic chemical transformations that result in three dominant forms of carbon-containing compounds: H_2CO_3 (carbonic acid), HCO_3^- (bicarbonate), and CO_3^{2-} (carbonate). Chemical equilibrium among these three chemical species occurs rapidly and leads to the following relations among their concentrations:

$$\frac{[H^+][HCO_3^-]}{[H_2CO_3]} = K_1 \tag{2}$$

$$\frac{[H^+][CO_3^{2-}]}{[HCO_3^-]} = K_2. \tag{3}$$

36. If this notation, the equation itself, and the chemistry of the carbonate complex (Eqs. 2 and 3 below) are unfamiliar, you may want to consult *COW-1*, pp. 104–8, for an introduction to this topic.

"Must be global warming, dude."

K_1 and K_2 are the dissociation constants for bicarbonate and carbonate. They, like Henry's constant, depend upon the temperature. Typical numerical values for K_1 and K_2 are $10^{-6.35}$ and $10^{-10.33}$, respectively. As we shall see, the temperature dependence of all three of these constants affects the stability of the oceans with respect to runaway outgassing of carbon dioxide.

It will be useful to define a new variable that describes the sum of these three concentrations of dissolved inorganic carbon, [DIC], in seawater:

$$[\mathrm{DIC}] = [\mathrm{H_2CO_3}] + [\mathrm{HCO_3}^-] + [\mathrm{CO_3}^{2-}]. \tag{4}$$

We can then calculate the total amount (as opposed to concentration) of DIC in the sea, in all three of these forms, by multiplying [DIC] by the total number of liters of seawater, N_S. Denoting the total amount of DIC by C_S, we get

$$C_S = N_S[\mathrm{DIC}]. \tag{5}$$

In the atmosphere, the only form of DIC is carbon dioxide itself. The total amount of DIC in the atmosphere, C_A, is the partial pressure, $p(CO_2)$, times the number of moles of atmosphere. Denoting the number of moles of atmosphere by N_A, we have

$$C_A = N_A p(CO_2). \tag{6}$$

We can now write the mass balance constraint on DIC as

$$C_T = C_A + C_S = N_A p(CO_2) + N_S[DIC]. \tag{7}$$

C_T is the total amount of DIC in the combined ocean–atmosphere systems. We will assume hereafter that C_T is a fixed constant, neglecting the possibility of net uptake or loss by rocks, soil, and living organisms.

To see if the positive feedback could lead to a sizable release of CO_2, we have to calculate a feedback factor that incorporates the effect of a change in surface temperature, T_S, on ocean carbon, the effect of a change in ocean carbon on atmospheric carbon, and the effect of a change in atmospheric carbon on surface temperature. In other words, our feedback factor will look like

$$g = \frac{\partial T_S}{\partial C_A} \frac{\partial C_A}{\partial C_S} \frac{\partial C_S}{\partial T_S}. \tag{8}$$

The first of these partial derivatives is just the greenhouse effect, which can be estimated from a variety of approaches, including simply taking output from general circulation models. In particular, current models project that surface temperature will increase by about 3 kelvin under a doubling of atmospheric CO_2. The preindustrial level of C_A was about 600 Gt(C), or 50 Pm(C), where the abbreviation Pm refers to petamoles, or 10^{15} moles. So a doubling of C_A brings the amount up by an increment of 50 Pm(C). Hence, as a very crude first approximation, we might assume

$$\begin{aligned} \frac{\partial T_S}{\partial C_A} &= \frac{3\ \text{K}}{50\ \text{Pm(C)}} \\ &= 6 \times 10^{-17}\ \text{K/mole(C)}. \end{aligned} \tag{9}$$

This is a reasonable estimate only for the first doubling of C_A. If all the oceanic DIC were to outgas to the atmosphere, so that C_A increased 50-fold, then the increase in T_S would not be $50 \times 3°C$ because of saturation effects. These effects are discussed in *COW-1*, where I derive an approximate numerical expression that includes these saturation effects.

What about the second partial derivative on the right side of Eq. 8? All the carbon outgassed from the sea goes to the atmosphere, so if we are careful to express both C_S and C_A in the same units, say petamoles of carbon, then this derivative is just equal to −1:

$$\frac{\partial C_A}{\partial C_S} = -1. \qquad (10)$$

All we have to do now is evaluate the third of the derivatives on the right side of Eq. 8. C_S is determined by [DIC], so let's rewrite [DIC] in a form that will allow us to make use of empirical information about the temperature dependence of $K_{H,CO2}$, K_1, and K_2. Using Eqs. 1 through 3, we can write

$$[DIC] = K_{H,CO2}\,p(CO_2)\left\{1 + \frac{K_1}{[H^+]} + \frac{K_1K_2}{[H^+]^2}\right\}, \qquad (11)$$

where the three terms in the braces correspond to carbonic acid, bicarbonate, and carbonate, respectively.

At this stage, it might appear that we can simply substitute in the temperature dependence of the K values in Eq. 10 and then use Eq. 5 to calculate the value of $\partial C_S/\partial T_S$. The problem with this strategy is that the term $[H^+]$ appears in Eq. 11, and this quantity will not remain constant when carbon dioxide enters or leaves the oceans.[37] We either have to eliminate somehow the term $[H^+]$ in Eq. 11, and then take the temperature derivative, or we have to include another chain of terms in Eq. 8 to account for $\partial[H^+]/\partial C_S$.

The former is much easier and more enlightening to perform, but to do so, we must review a little more aquatic chemistry. Although $[H^+]$ changes, another chemical quantity does remain approximately constant when CO_2 enters or leaves an aqueous solution. That quantity is called alkalinity, which is a measure of the ability of the solution to resist changes in its pH. Denoted by [Alk], the concentration of alkalinity in seawater is approximately given by the expression

$$[Alk] = [HCO_3{}^-] + 2[CO_3{}^{2-}] + [OH^-] - [H^+]. \qquad (12)$$

From the very form of Eq. 12, it is clear that when CO_2 is added to water, this [Alk] does not change. Suppose a mole of CO_2 in the form of H_2CO_3 dissociates to a mole of HCO_3^- and a mole of H^+. These dissociation products appear in Eq. 12 with opposite signs and in the same proportion, and so each mole of H^+ is balanced by a mole of

37. Eqs. 2 and 3 should suggest to you that when carbon dioxide either enters or leaves the oceans, the pH of the seawater will change. The pH of a solution is defined by $pH = -\log_{10}[H^+]$. Because $[H^+]$ occurs in the equilibrium relationships in Eqs. 2 and 3, the pH is likely to be altered when a new equilibrium is achieved. So we cannot assume in our estimation of $\partial C_S/\partial T_S$ that pH is constant.

HCO_3^-; no net change in alkalinity occurs. Moreover, when a mole of HCO_3^- dissociates to a mole of CO_3^{2-} plus a mole of H^+, then one mole of HCO_3^- is lost, and two moles of H^+ and one mole of CO_3^{2-} are formed; because of the factor of 2 in front of the $[CO_3^{2-}]$ term in Eq. 12, [Alk] again remains unchanged.[38]

Our approach is to eliminate $[H^+]$ from Eq. 11 by replacing it with a function of the constant [Alk]. To express $[H^+]$ as a function of alkalinity we rewrite Eq. 12 in the following form:

$$[\mathrm{Alk}] = K_{\mathrm{H,CO2}}\mathrm{p}(\mathrm{CO_2})\left\{\frac{K_1}{[\mathrm{H^+}]} + \frac{2K_1K_2}{[\mathrm{H^+}]^2}\right\} + \frac{K_\mathrm{w}}{[\mathrm{H^+}]} - [\mathrm{H^+}]. \tag{13}$$

Here, we have used Eqs. 1 through 3 and replaced $[OH^-]$ with its equilibrium value, where $K_w = 10^{-14}$ is the dissociation constant for water. We could, in principle, solve this equation for $[H^+]$ as a function of [Alk], but because that process involves a cubic equation, the result is rather messy. When that result is then substituted into Eq. 11, the resulting expression is messier still. So let's simplify life by taking advantage of the fact that in the ocean, some of the terms in Eqs. 11 and 13 are much smaller than others.

The pH of seawater varies with depth and from ocean to ocean but averages about 8.2; in other words, $[H^+] \sim 10^{-8.2}$. In seawater today, the magnitudes of the three terms in braces on the right side of Eq. 11 are roughly in the ratios: 1:150:7.5. In other words, the bicarbonate term is 150 times larger than the carbonic acid term and about 20 times larger than the carbonate term. Similarly, in Eq. 13, the bicarbonate and carbonate terms are the dominant ones, and the former is about 10 times larger than the latter. Keeping only the bicarbonate and carbonate terms in Eq. 13, and solving for $[H^+]$, we get

$$[\mathrm{H^+}] = \frac{1}{2}\left\{\frac{K_1K_\mathrm{H}\mathrm{p}(\mathrm{CO_2})}{[\mathrm{Alk}]}\right\}\left\{1 + \left[1 + \frac{8K_2[\mathrm{Alk}]}{\mathrm{p}(\mathrm{CO_2})K_\mathrm{H}K_1}\right]^{1/2}\right\}. \tag{14}$$

Here, the term $8K_2[\mathrm{Alk}]/(\mathrm{p}(\mathrm{CO_2})K_\mathrm{H}K_1)$, which arose from the carbonate contribution to alkalinity, is ~ 0.15. Because this is considerably smaller than 1, we can use the approximation $(1+\varepsilon)^{1/2} \sim 1 + \varepsilon/2$, which is valid if $\varepsilon \ll 1$, to arrive at the simpler expression,

$$[\mathrm{H^+}] \cong \left\{\frac{K_1K_\mathrm{H}\mathrm{p}(\mathrm{CO_2})}{[\mathrm{Alk}]}\right\}\left\{1 + \frac{2K_2[\mathrm{Alk}]}{\mathrm{p}(\mathrm{CO_2})K_\mathrm{H}K_1}\right\}$$

38. Alkalinity can change if, say, the reduced pH from CO_2 addition causes the dissolution of seashells or other materials containing solid calcium carbonate. This effect occurs in seawater, but to a first approximation, it can be neglected in the present calculation. I thank Steve Fetter for suggesting the approximations that take us from Eq. 11 to Eq. 16.

$$= \frac{K_1 K_H p(CO_2)}{[Alk]} + 2K_2. \tag{15}$$

Substituting this expression for $[H^+]$ into Eq. 11 but again treating the bicarbonate term exactly, expanding in the second-order terms arising from the carbonate contribution, and neglecting the carbonic acid term in Eq. 11, we get

$$[DIC] \cong K_{H,CO2} p(CO_2) \left\{ \frac{K_1}{[H^+]} + \frac{K_1 K_2}{[H^+]^2} \right\}$$

$$\cong [Alk] + \frac{(K_2/K_1 K_H)[Alk]^2}{p(CO_2)}. \tag{16}$$

Substituting this expression into Eq. 5, we get

$$C_S = N_S \left\{ [Alk] + \frac{(K_2/K_1 K_H)[Alk]^2}{p(CO_2)} \right\}. \tag{17}$$

To take the partial derivative of this equation with respect to T_S, we note that only the combination $(K_2/K_1 K_H)$ is temperature-dependent.[39] Thus, we get

$$\frac{\partial C_S}{\partial T_S} = N_S \left\{ \frac{\partial (K_2/K_1 K_H)}{\partial T_S} \right\} \frac{[Alk]^2}{p(CO_2)}. \tag{18}$$

To evaluate this expression numerically, we have to determine the temperature dependence of the particular combination of constants, $\beta = K_2/K_1 K_H$, that appears in Eq. 18. The following table provides empirical values for β as a function of temperature:

T(K)	β
273	114.7
278	90.6
283	71.1
288	57.9
293	48.2
298	36.1
303	28.1
308	21.8

39. Actually, $p(CO_2)$ is temperature-dependent because of a number of biological, biogeochemical, and physical processes, but within the set of processes considered here, the dependence of $p(CO_2)$ on temperature is entirely the result of the fact that as the ocean outgasses, CO_2 passes from the ocean to the atmosphere, and this effect is already captured in Eq. 8 by the term $(\partial C_A/\partial C_S)$.

Ambient sea water temperature, averaged over all the oceans, is about 275 K, whereas that of the mixed or surface layer (the top ~100 m) is about 295 K. If we assume the entire ocean takes part in the feedback process, then, reading from the table, we have

$$\frac{\partial \beta}{\partial T_S} \sim \frac{114.7 - 90.6}{273 - 278} = -4.82. \tag{19}$$

The units of $\partial\beta/\partial T_S$ are $(K_H)^{-1}(\text{kelvins})^{-1}$ = (liters)(atmospheres)/(mole)(kelvins). If we assume that only the mixed layer of the ocean takes part in the feedback process, then

$$\frac{\partial \beta}{\partial T_S} \sim \frac{48.2 - 36.1}{293 - 298} = -2.42. \tag{20}$$

In the former case, we would take $N_S = 1.35 \times 10^{15}$ liters, and in the latter case, $N_S = 2.7 \times 10^{13}$ liters. Exercise 3 asks you to think about the circumstances under which you would choose the former or the latter approach.

We can now evaluate Eq. 8 by combining the results in Eqs. 9, 10, 18, and either 19 or 20:

$$g = [6 \times 10^{-17}] \cdot [-1] \cdot \left[N_S \left\{ \frac{\partial(K_2/K_1K_H)}{\partial T_S} \right\} \frac{[\text{Alk}]^2}{\text{p}(CO_2)} \right]. \tag{21}$$

The numerical value for [Alk] is $\sim 2.45 \times 10^{-3}$ moles/liter and that for $\text{p}(CO_2)$ is 365×10^{-6} atmospheres. Assuming the entire ocean takes part in the feedback, so that $N_S = 2.7 \times 10^{13}$ liters, we get

$$g = [6 \times 10^{-17}] \times [-1] \times \frac{(1.35 \times 10^{15})(-4.82)(2.45 \times 10^{-3})}{3.65 \times 10^{-4}} = +0.0064. \tag{22}$$

The feedback is indeed positive, but it is most definitely less than 1, indicating that there is no instability; we need not worry about our ocean going flat as a consequence of climate warming!

EXERCISE 1: By inserting the appropriate units of each of the terms that we multiply together to form g in Eq. 21, work out the units of g. Does your answer make sense?

EXERCISE 2: Evaluate the feedback factor, g, if only the mixed layer takes part in the feedback process.

EXERCISE 3: What does it mean to say that "the entire ocean (or only the mixed layer) takes part in the feedback process?" Under what conditions would you make one or the other of these assumptions?

EXERCISE 4: During the past 100 years, the average surface temperature of the earth has warmed about 0.6 kelvin. Assuming that such a warming characterizes the mixed layers of all the oceans, by how much should that warming have increased the atmospheric level of carbon dioxide via the mechanism discussed here?

EXERCISE 5: What would have to be different about our planet for the feedback factor, g, to be ≥ 1, and thus for Earth to undergo a runaway greenhouse effect? Consider such factors as the size or ambient temperature of Earth's oceans, albedo, distance from the Sun, $p(CO_2)$, etc. Lest this seem like merely an academic puzzle, note that it has been suggested that a runaway greenhouse effect might have occurred long ago on the planet Venus, which would account for its current atmosphere comprised nearly entirely of carbon dioxide, its surface temperature of $\sim$750 K, and the absence of water on its surface.

Appendix
Useful Mathematical Expressions and Approximations

1. Notations

$\sum_{i=1}^{N} a_i = a_1 + a_2 + \cdots + a_N$

$\ln(x)$ = logarithm of x to base $e = 2.718\ldots$

$\log_a(x)$ = log of x to the base a (that is, $a^{\log_a(x)} = x$)

$n! = n(n-1)(n-2)(n-3)\ldots(2)(1) \equiv$ "n factorial" (n must be an integer)

$\binom{N}{M} = \frac{N!}{(M!)(N-M)!}$

$\langle X \rangle = \bar{X}$ = average of the X values

$\sigma^2(X)$ = variance of the X values $= \langle X^2 \rangle - \langle X \rangle^2$

V_i = elements of the vector **V**: $\mathbf{V} = (V_1, V_2, \ldots)$

M_{ij} = elements of the matrix **M**; the first index, i, labels rows, the second labels columns

Thus, if **M** is an $N \times N$ matrix:

$$\mathbf{M} = \begin{bmatrix} M_{11} & M_{12} & \ldots & M_{1N} \\ M_{21} & M_{22} & \ldots & M_{2N} \\ \cdot & \cdot & & \\ M_{N1} & M_{N2} & \ldots & M_{NN} \end{bmatrix}$$

"Sum-over-repeated-index rule":

$$M_{ij}V_j = \sum_j M_{ij}V_j = M_{i1}V_1 + M_{i2}V_2 + \cdots$$

$$M_{ij}Q_{jk} = M_{i1}Q_{1k} + M_{i2}Q_{2k} + \cdots$$

I = unit matrix: $I_{ii} = 1$; $I_{ij} = 0$ if $i \neq j$ (these elements of **I** are sometimes denoted by the kronecker-delta symbol δ_{ij})

$|\mathbf{M}|$ = determinant of **M**

2. Geometric Formulas

Areas

Triangle: $\frac{\text{(base)(height)}}{2}$

Circle: $\pi(\text{radius})^2$

Sphere: $4\pi(\text{radius})^2$

Volumes

Sphere: $\left(\frac{4\pi}{3}\right)(\text{radius})^3$

Cone: $(\frac{1}{3})$(area of base)(height)

3. Trigonometric Identities

$\cos^2 x + \sin^2 x = 1$

$\sin(x + y) = (\sin x)(\cos y) + (\cos x)(\sin y)$

$\cos(x + y) = (\cos x)(\cos y) - (\sin x)(\sin y)$

4. Logarithms

$\log_a(xy) = \log_a(x) + \log_a(y)$

$\log_a \frac{x}{y} = \log_a(x) - \log_a(y)$

$\log_a(x^y) = y \log_a(x)$

$\log_b(x) = \log_b(a) \log_a(x)$

$\log_b(a) = \frac{1}{\log_a(b)}$

$\log_e(2) = 0.693$

$\log_e(10) = 2.303$

$\log_{10}(2) = 0.301$

5. Complex Numbers ($i = \sqrt{-1}$)

$z = a + bi = \rho e^{i\theta} = \rho \cos\theta + i\rho \sin\theta$, where $\rho^2 = a^2 + b^2$, and $\theta = \tan^{-1}\frac{b}{a}$ (note that θ must be expressed in radians)

A wonderful formula: $e^{2\pi i} + 1 = 0$

6. Derivatives

$$\frac{d(x^a)}{dx} = ax^{a-1}$$

$$\frac{d[\sin f(x)]}{dx} = \frac{df(x)}{dx} \cos[f(x)]$$

$$\frac{d[\cos f(x)]}{dx} = -\frac{df(x)}{dx} \sin[f(x)]$$

$$\frac{d(e^{f(x)})}{dx} = \frac{df(x)}{dx} e^{f(x)}$$

$$\frac{d[\ln f(x)]}{dx} = \frac{df(x)/dx}{f(x)}$$

7. Integrals

$$\int x^a \, dx = (a+1)^{-1} x^{a+1} \quad a \neq -1$$

$$\int \frac{1}{x} dx = \ln(x)$$

$$\int a^x \, dx = [\ln(a)]^{-1} a^x$$

$$\int \frac{1}{x^2 + a^2} \, dx = a^{-1} \tan^{-1} \frac{x}{a}$$

$$\int \sin(x) \, dx = -\cos(x)$$

$$\int \cos(x) \, dx = \sin(x)$$

$$\int \sin^2 x \, dx = \frac{x}{2} - \frac{\sin(2x)}{4}$$

$$\int \tan(x) \, dx = -\ln[\cos(x)]$$

8. Summations

$$\sum_{n=0}^{N} a^n = \frac{a^{N+1}-1}{a-1}$$

For $-1 < a < 1$, $\sum_{n=0}^{\infty} a^n = \frac{1}{1-a}$

$$\sum_{n=0}^{N} na^n = \frac{a}{a-1}\left[(N+1)a^N - \frac{a^{N+1}-1}{a-1}\right]$$

For $-1 < a < 1$, $\sum_{n=0}^{\infty} na^n = \frac{a}{(1-a)^2}$

$$\sum_{n=1}^{N} n = \frac{N(N+1)}{2}$$

$$\sum_{n=1}^{N} n^2 = \frac{(N)(N+1)(2N+1)}{6}$$

9. Approximations

Note: In the following equations, the inequality $|\varepsilon| < 1$ is assumed to hold, where $|\;|$ means "absolute value."

The formulas below are all special cases of the general Taylor series expansion:

$$f(\varepsilon) = \sum_{n=0}^{\infty} \frac{\varepsilon^n}{n!} \frac{d^{(n)}f(x)}{dx^{(n)}}\bigg|_{x=0} = f(0) + \frac{\varepsilon}{1!}\frac{df(x)}{dx}\bigg|_{x=0} + \frac{\varepsilon^2}{2!}\frac{d^2f(x)}{dx^2}\bigg|_{x=0} + \cdots$$

$$\sim f(0) + \frac{\varepsilon\, df(x)}{dx}\bigg|_{x=0}$$

$$\frac{1}{1+\varepsilon} = 1 - \varepsilon + \varepsilon^2 + \cdots \sim 1 - \varepsilon$$

$$\frac{1}{1-\varepsilon} = 1 + \varepsilon + \varepsilon^2 + \cdots \sim 1 + \varepsilon$$

$$(1+\varepsilon)^{1/2} = 1 + \frac{\varepsilon}{2} - \frac{\varepsilon^2}{8} + \cdots \sim 1 + \frac{\varepsilon}{2}$$

More generally: $(1+\varepsilon)^a = 1 + a\varepsilon + \frac{a(a-1)\varepsilon}{2} + \frac{a(a-1)(a-2)\varepsilon^3}{6} + \cdots \sim 1 + a\varepsilon$

$$e^{\varepsilon} = \sum_{n=0}^{\infty} \frac{\varepsilon^n}{n!} = 1 + \varepsilon + \frac{\varepsilon^2}{2} + \frac{\varepsilon^3}{6} + \cdots \sim 1 + \varepsilon$$

$$\sin(\varepsilon) = \varepsilon - \frac{\varepsilon^3}{6} + \cdots \sim \varepsilon$$

$$\cos(\varepsilon) = 1 - \frac{\varepsilon^2}{2!} + \frac{\varepsilon^4}{4!} \cdots \sim 1 - \frac{\varepsilon^2}{2}$$

$$\ln(1+\varepsilon) = \varepsilon - \frac{\varepsilon^2}{2} + \frac{\varepsilon^3}{3} - \frac{\varepsilon^4}{4} + \cdots \sim \varepsilon$$

$(1+\varepsilon)^{a/\varepsilon} \to e^a$ as $\varepsilon \to 0$

$\sigma^2[f(x)] \sim \left(\frac{df}{dx}\right)^2 \sigma^2(x)$ [valid to the extent $f(x)$ is slowly varying around $x = \langle x \rangle$]

Stirling's approximation: $x! \sim (2\pi x)^{1/2} x^x e^{-x}$ (for large x)

10. Vectors and Matrices

Matrix algebra

Product of matrix times vector: $(\mathbf{AV})_i = A_{ij}V_j$ (use the repeated index rule to evaluate this!)

Product of matrix times matrix: $(\mathbf{AQ})_{ik} = A_{ij}Q_{jk}$ (use the repeated index rule to evaluate this!)

Determinants

Let $\mathbf{M}_{(ij)}$ be the matrix obtained by striking out the ith row and j^{th} column of $\mathbf{A}$. Then the determinant of $\mathbf{A} = \sum$ determinant $(\mathbf{M}_{(ij)})A_{ij}(-1)^{i+j}$. (This expression can be evaluated for any fixed value of the index i.)

The determinant of $\mathbf{M}_{(ij)}$ is called the "minor" of the matrix element A_{ij}.

The expression $(-1)^{i+j}$ times the determinant of $\mathbf{M}_{(ij)}$ is called the "cofactor" of the matrix element A_{ij}.

The inverse of a matrix $\mathbf{A}$ has matrix elements given by $(A^{-1})_{ij} = M_{ji}/\text{determinant}\ (\mathbf{A})$, where M_{ji} is a cofactor of A_{ji}.

Eigenvalues and eigenvectors

The eigenvalues, λ, and the eigenvectors, $\mathbf{V}$, of a matrix $\mathbf{A}$ satisfy $(\mathbf{A} - \lambda\mathbf{I})\mathbf{V} = 0$, which implies that the determinant of $(\mathbf{A} - \lambda\mathbf{I}) = 0$.

Example of an eigenvalue–eigenvector calculation:

$$\text{Let } \mathbf{A} = \begin{bmatrix} 2 & 1 \\ -5 & 0 \end{bmatrix}.$$

Then the eigenvalues are determined by setting the determinant of

$$\mathbf{A} - \lambda\mathbf{I} = \begin{bmatrix} 2-\lambda & 1 \\ -5 & -\lambda \end{bmatrix}$$

equal to zero. This leads to the "characteristic equation" for the eigenvalues:

$$\lambda^2 - 2\lambda + 5 = 0.$$

The solutions to this equation are $\lambda_1 = 1 + 2i$ and $\lambda_2 = 1 - 2i$.

Having obtained the two eigenvalues, we then determine the associated eigenvectors $\mathbf{V_1}$ and $\mathbf{V_2}$ by $\mathbf{AV_1} = \lambda_1\mathbf{V_1}$ and $\mathbf{AV_2} = \lambda_2\mathbf{V_2}$. Consider $\mathbf{V_1}$ and let the two components of the vector be $\mathbf{V}_{1,1}$ and $\mathbf{V}_{1,2}$. Then:

$$\begin{bmatrix} 2 & 1 & \mathbf{V}_{1,1} \\ -5 & 0 & \mathbf{V}_{1,2} \end{bmatrix} = \lambda_1 \begin{bmatrix} \mathbf{V}_{1,1} \\ \mathbf{V}_{1,2} \end{bmatrix}$$

Using the rule for multiplying a matrix times a vector, we obtain two equations:

$$2\mathbf{V}_{1,1} + \mathbf{V}_{1,2} = \lambda_1\mathbf{V}_{1,1}$$

and

$$-5\mathbf{V}_{1,1} = \lambda_1\mathbf{V}_{1,2}.$$

Using $\lambda_1 = 1 + 2i$, we find the first of these equations tells us that $\mathbf{V}_{1,1}/\mathbf{V}_{1,2} = -1/(1-2i)$. The second equation is consistent with the first and provides no new information. Thus, all we can determine is the ratio of the two elements of the eigenvector. This means we can write

$$\mathbf{V_1} = \begin{bmatrix} -\dfrac{c}{1-2i} \\ c \end{bmatrix},$$

where c is an arbitrary constant.

For the second eigenvector, $\mathbf{V_2}$, a similar expression is obtained: $\mathbf{V_{2,1}}/\mathbf{V_{2,2}} = -1/(1+2i)$, which means we can write

$$\mathbf{V_2} = \begin{bmatrix} -\dfrac{d}{1+2i} \\ d \end{bmatrix}$$

where d is an arbitrary constant.

The matrix $\mathbf{C}$ in Eq. 13 from the Background section of Chapter V is then given by

$$\mathbf{C} = \begin{bmatrix} -\dfrac{c}{1-2i} & -\dfrac{d}{1+2i} \\ c & d \end{bmatrix}.$$

Because of the presence of the two undetermined constants c and d, this result might appear to yield an ambiguous expression for the behavior of the perturbed system variables (Eq. 13 in Background, Chapter V). In fact, because of the presence of the pair of matrices $\mathbf{C}$ and $\mathbf{C}^{-1}$ in Eq. 13, these constants disappear in the final expression for the $\Delta X_i(t)$.

Because the eigenvalues and eigenvectors are complex, it also might appear that Eq. 13 would predict complex values for the $\Delta X_i(t)$. This would be a problem because the $\Delta X_i(t)$, being measurable entities (number of rabbits and foxes, for example), must be described by real, not complex, numbers. In fact, because the eigenvalues are complex conjugates of each other (they differ only in the sign of the term containing i), the calculated values of the $\Delta X_i(t)$ come out to be real.

Further Reading

General Mathematics

Abramowitz, M., Stegun, I. 1972. *Handbook of mathematical functions: with formulas, graphs, and mathematical tables*. New York: Dover Press.

Boyce, W., DiPrima, R. 1986. *Elementary differential equations*. New York: Wiley.

Logan, J. 1987. *Applied mathematics: a contemporary approach*. New York: Wiley.

Swartz, C. E. 1993. *Used math: for the first two years of college science*. College Park, MD: AAPT Press.

Probability

Evans, M., Hastings, N., Peacock, B. 1993. *Statistical distributions*. New York: Wiley.

Feller, W. 1968. *An introduction to probability theory and its applications*, 3rd ed. New York: Wiley.

Kammen, D., Hassenzahl, D. M. 1999. *Should we risk it? Exploring environmental, health and technological problem solving*. Princeton, NJ: Princeton Univ. Press.

Taylor, J. R. 1997. *An introduction to error analysis*, 2nd ed. Sausalito, Calif.: University Science Books.

Optimization

Beltrami, E. 1998. *Mathematics for dynamic modeling*, 2nd ed. Boston: Academic Press.

Clark, C. 1976. *Mathematical bioeconomics, the optimal control of renewable resources*. New York: Wiley.

Costanza, R., ed. 1991. *Ecological economics: the science and management of sustainability*. New York: Columbia Univ. Press.

Mangel, M., Clark, C. 1988. *Dynamic modeling in behavioral ecology*. Princeton, NJ: Princeton Univ. Press.

Scaling

Barenblatt, G. 1996. *Scaling, self similarity, and intermediate asymptotics*. Cambridge: Cambridge Univ. Press.

Rodriguez-Iturbe, I., Rinaldo, A. 1997. *Fractal river basins: chance and self-organization in hydrology*. Cambridge: Cambridge Univ. Press.

Rosenzweig, M. 1995. *Species diversity in space and time*. Cambridge: Cambridge Univ. Press.

Schroeder, M. 1991. *Fractals, chaos, and power laws: minutes from an infinite paradise*. New York: Freeman.

Sposito, G. 1998. *Scale dependence and scale invariance in hydrology*. Cambridge: Cambridge Univ. Press.

Differential Equations

Box models, general

Ford, A. 1999. *Modeling the environment*. Washington, D.C.: Island Press.

Harte, J. 1988. *Consider a spherical cow: a course in environmental problem solving*. Sausalito, Calif.: University Science Press.

Biogeochemistry

Broecker, W. 1974. *Chemical oceanography*. New York: Harcourt Brace Jovanovich.

Garrels, R., Mackenzie, F., Hunt, C. 1975. *Chemical cycles and the global environment: assessing human influences*. Los Altos, Calif.: Kaufmann, Inc. (out of print but in most university libraries).

Schlesinger, W. H. 1991. *Biogeochemistry: an analysis of global change*. New York: Academic Press.

Population biology and ecology

May, R. 1975. *Stability and complexity in model ecosystems*. Princeton, NJ: Princeton Univ. Press.

Murray, J. 1989. *Mathematical biology*. New York: Springer-Verlag.

Roughgarden, J. 1998. *Primer of theoretical ecology*. New York: Prentice Hall.

Climate

Gethner, R. 1998. Climate change and the daily temperature cycle. *The UMAP Journal* 19(1):33–86.

Henderson-Sellers, A., McGuffie, K. 1987. *A climate modeling primer*. New York: Wiley.

Houghton, J. 1977. *The physics of atmospheres*. Cambridge: Cambridge Univ. Press.

Stability and Feedback

LaSalle, J., Lefshetz, S. 1960. *Stability by Liapunov's direct method*. New York: Academic Press.

May, R. 1975. *Stability and complexity in model ecosystems*. Princeton, NJ: Princeton Univ. Press.

Useful Sources of Data

Gleick, P., ed. 1993. *Water in crisis: a guide to the world's fresh water resources*. New York: Oxford Univ. Press.

Harte, J., Holdren, C., Schneider, R., Shirley, C. 1990. *Toxics a to z: a guide to everyday pollution hazards*. Berkeley, Calif.: Univ. California Press.

Intergovernmental Panel on Climate Change. 1996. *Climate change: the IPCC scientific assessment*. Cambridge: Cambridge Univ. Press.

Turner, B., Clark, W., Kates, R., Richards, J., Mathews, J., Meyer, W., eds. 1991. *The Earth as transformed by human action: global and regional changes in the biosphere over the past 300 years*. Cambridge: Cambridge Univ. Press.

Weast, R. C. 1992. *Handbook of chemistry and physics: a ready-reference book of chemical and physical data*, 73rd edition. Boca Raton, Fla.: CRC Press.

World Resources Institute, 1999. *World Resources 1998–99*. New York: Oxford Univ. Press.

Index